Noncommutative Mathematics for Quantum Systems

Noncommutative mathematics is a significant new trend of mathematics in the twentieth century. Initially motivated by the formulation and development of quantum physics due to Heisenberg and von Neumann, the idea of 'making a theory noncommutative' has been extended to many areas of pure and applied mathematics. An example is quantum probability, describing the probabilistic aspects of quantum mechanics. The generalization from classical to quantum happens here in two steps: first the theory is reformulated in terms of algebras of functions on probability spaces, then the commutativity condition is dropped.

This book focuses on two current areas of noncommutative mathematics: quantum probability and quantum dynamical systems.

The first part of the book provides an introduction to quantum probability and quantum Lévy processes. It provides introduction to the notion of independence in quantum probability. The theory of quantum stochastic processes with independent and stationary increments is also highlighted. The second part provides an introduction to quantum dynamical systems. It focuses on analogies with fundamental problems studied in classical dynamics.

The text underlines the balance between two crucial aspects of noncommutative mathematics. On one hand the desire to build an extension of the classical theory provides new, original ways to understand well-known 'commutative' results and on the other hand the richness of the quantum mathematical world presents completely novel phenomena, never encountered in the classical setting. This text will be useful to students and researchers in noncommutative probability, mathematical physics and operator algebras.

Uwe Franz is Professor at the University of Franche-Comté in Besançon (France) since 2005. He authored the book *Stochastic Processes and Operator Calculus on Quantum Groups*, edited four books on quantum probability and has written over 50 peer-reviewed research papers in the area of noncommutative mathematics. His areas of interest include noncommutative probability, quantum stochastic processes, quantum stochastic calculus and probability on quantum groups.

Adam Skalski is Associate Professor at the Institute of Mathematics, Polish Academy of Sciences and at the University of Warsaw, Poland. He has been awarded the Kuratowski Prize of the Polish Mathematical Society in 2008 and the Sierpiński Prize of the Polish Academy of Sciences in 2014. His areas of interest include topological quantum groups, operator algebras and quantum stochastic processes.

CAMBRIDGE–IISc SERIES

Cambridge–IISc Series aims to publish the best research and scholarly work on different areas of science and technology with emphasis on cutting-edge research.

The books will be aimed at a wide audience including students, researchers, academicians and professionals and will be published under three categories: research monographs, centenary lectures and lecture notes.

The editorial board has been constituted with experts from a range of disciplines in diverse fields of engineering, science and technology from the Indian Institute of Science, Bangalore.

Cambridge - IISc Series

Noncommutative Mathematics for Quantum Systems

Uwe Franz
Adam Skalski

CAMBRIDGE
UNIVERSITY PRESS

CAMBRIDGE
UNIVERSITY PRESS

4843/24, 2nd Floor, Ansari Road, Daryaganj, Delhi - 110002, India

Cambridge University Press is part of the University of Cambridge.

It furthers the University's mission by disseminating knowledge in the pursuit of education, learning and research at the highest international levels of excellence.

www.cambridge.org
Information on this title: www.cambridge.org/9781107148055

First published 2016

Printed in India

A catalogue record for this publication is available from the British Library

Library of Congress Cataloguing-in-Publication data
Franz, Uwe.
Noncommutative mathematics for quantum systems / Uwe Franz, Adam Skalski.
pages cm
Includes bibliographical references and index.
Summary: "Discusses two current areas of noncommutative mathematics, quantum probability and quantum dynamical systems"– Provided by publisher.
ISBN 978-1-107-14805-5 (hardback)
1. Probabilities. 2. Quantum theory. 3. Potential theory (Mathematics) I. Skalski, Adam, 1978- II. Title.
QC174.17.P68F73 2016
530.13'3–dc23
2015032903

ISBN 978-1-107-14805-5 Hardback

We would like to dedicate this book to Marysia,
who was 11 days old when the Bangalore school began.

Contents

Preface

This monograph arose from the lectures delivered by the authors during the graduate school, being a part of the meeting 'Recent advances in Operator Theory and Operator Algebras', which took place between 31 December 2012 and 12 January 2013 in the Indian Statistical Institute in Bangalore. We are very grateful to Rajarama Bhat, Tirthankar Bhattacharyya and Jaydeb Sarkar for organizing the meeting, inviting us to speak and providing excellent working conditions during our stay at ISI, and the audience of the school for their active participation in our courses. The authors are strong believers in the usefulness of organizing graduate schools for mathematical students – in fact we first met during an analogous event in Greifswald (Germany) in 2003, respectively, as a lecturer and a participating PhD student, and have collaborated ever since.

Our lectures during the Bangalore school treated notions of independence and quantum Lévy processes in quantum probability (by the first named author) and quantum processes understood as noncommutative incarnations of classical dynamical systems (by the second named author). Both these topics, born from necessity to incorporate quantum models into mathematical approaches to study physical systems, have now become very active, broad areas of modern mathematical research. The monograph consists of two chapters: 'Independence and Lévy processes in quantum probability', authored by Uwe Franz, and 'Quantum dynamical systems from the point of view of noncommutative mathematics', authored by Adam Skalski. These can be read independently, but we believe that there is an added value in placing them together; not in the least because they present alternative approaches to the noncommutative/quantum generalizations of classical concepts and results. The monograph is essentially self-contained, with several references to both sources of necessary background and to current research literature, and should thus form an appropriate entrance point for graduate students interested in the area.

We would like to express our gratitude to Rajarama Bhat and Gadadhar Misra for encouraging us to transform the lecture notes into the monograph form, and facilitating the contacts with the publisher, Cambridge University Press India, whose assistance during the editing process is also gratefully acknowledged.

We are indebted to Takahiro Hasebe, Anna Kula, Souleiman Omar Hoche, Stephanie Lachs, Michaël Ulrich, Simeng Wang, and Mateusz Wasilewski for their comments, suggestions, and corrections of an earlier version of these notes. Special thanks are due to Sergey Neshveyev and Michael Schürmann for their critical reading of the early draft of the text and helpful remarks.

While the book was being written, UF was supported by the ANR Project OSQPI (ANR-11-BS01-0008), and AS was partially supported by the Iuventus Plus grant IP2012 043872. UF also wants to thank the Alfried Krupp Wissenschaftskolleg in Greifswald, where he spent the summer term 2014 as a research fellow, for its hospitality and for providing the ideal environment for revising these notes.

The Bangalore Center of the Indian Statistical Institute (ISI) has been holding a Workshop cum Conference on Operator Theory and Operator Algebras almost every alternate year for the last few decades. It has been growing in popularity. The name of the meeting has now stabilized to Recent Advances in Operator Theory and Operator Algebras (OTOA). More details about these workshops can be found at: http://www.isibang.ac.in/~statmath/conferences/conferences.html

In 2012–13, OTOA was organized by B.V. Rajarama Bhat, Tirthankar Bhattacharyya and Jaydeb Sarkar and was sponsored by ISI and the National Center for Mathematics (India). The main speakers in the workshop were Uwe Franz, K.R. Parthasarathy, Adam Skalski and Brett D. Wick.

Conference photo of the meeting
"Recent advances in Operator Theory and Operator Algebras 2012" ISI, Bangalore, Dec. 31, 2012 - Jan. 11, 2013

Introduction

Noncommutative mathematics might be considered as one of the major new trends of mathematics that arose in the twentieth century. Initially motivated by the formulation and development of quantum physics due to Heisenberg and von Neumann, the idea of 'making a theory noncommutative' has been since extended to many areas of pure and applied mathematics and turned out to be extremely fruitful.

In 1925, Heisenberg suggested replacing the observables of a quantum system, such as its position Q and momentum P, by noncommutative quantities, most commonly realised as operators on a Hilbert space. This idea allowed the correct prediction of the results of the experiments of quantum physics; for example, it led to a description of the observed spectra of atoms. It has also led to a cornucopia of new mathematics, in particular to the theory of von Neumann algebras and C^*-algebras developed, respectively, by Murray and von Neumann, and by Gelfand and Naimark.

The general pattern of the noncommutative mathematics is the following: take a 'classical' mathematical theory, say topology, measure theory, differential geometry, or group theory, and reformulate it in terms of algebras. These algebras are algebras of functions on the spaces appearing in the 'classical' theory, that is, continuous functions (with values in $\mathbb{C}$) on a topological space, measurable functions on a measure space, smooth functions on a manifold, and so on. They inherit some additional structure, for example, an involution, a norm or topology, a coproduct, from the space on which they are defined. Finally, these algebras are always commutative. Axiomatizing the additional structure and dropping the commutativity condition one arrives at a noncommutative generalization of the original theory. In this way the theory of von Neumann algebras or C^*-algebras can be viewed, respectively, as noncommutative measure theory and noncommutative topology. The theory of Hopf algebras can be viewed as noncommutative group theory (here a warning is in place: the term 'noncommutative' refers to the function algebra, not to the group multiplication, which can also be noncommutative in classical 'commutative' group theory).

The procedure of making a theory noncommutative is not canonical, the choice of the axioms of the new noncommutative theory is not unique and each time has to be adapted to the classical theory to which it is applied. Because the idea originated from quantum physics, this procedure is also referred to as quantization and the new theory is labeled a 'quantum' theory. A good noncommutative theory should allow extending the central results of the classical theory and it should contain the classical theory in a clear way. A good example for the latter is the Gelfand functor, which defines an equivalence between the category of locally compact topological spaces and the category of commutative C^*-algebras.

The two lectures in this volume aim to present the rich new mathematics that was discovered in this way. We will concentrate on two fields of noncommutative mathematics: quantum probability, presented in the first chapter of this monograph, 'Independence and Lévy processes in quantum probability' (written by Uwe Franz), and quantum dynamical systems, treated in the second chapter, 'Quantum dynamical systems from the point of view of noncommutative mathematics' (written by Adam Skalski).

Quantum probability is a generalization of both classical probability theory and quantum mechanics that allows to describe the probabilistic aspects of quantum mechanics. 'Classical' probability spaces $(\Omega, \mathcal{A}, P)$ are replaced by the pairs $(L^\infty(\Omega), E(\cdot) = \int_\Omega \cdot \mathrm{d}P)$ consisting of the commutative von Neumann algebra of bounded random variables and the expectation functional. Then, the commutativity condition is dropped. In this way we arrive at the notion of a (von Neumann) algebraic probability space (N, Φ) consisting of a von Neumann algebra N and a normal (faithful tracial) state Φ. As we have seen this includes classical probability spaces in the form $(L^\infty(\Omega), E)$, however, also quantum mechanical systems modeled by a Hilbert space H and a pure state $\psi \in H$ (or a mixed state $\rho \in S(H)$), if we take $N = B(H)$ and Φ the state defined by $\Phi(X) = \langle \psi, X\psi \rangle$ (or $\Phi(X) = \mathrm{tr}(\rho X)$ for $X \in B(H)$).

We explain the setting of quantum probability in more detail in the second section of Franz' lecture. In the third section we discuss a version of the EPR experiment and a theorem by Kochen and Specker to explain that physics probably forces us to use quantum probability to describe our world at the quantum level.

Independence and Lévy processes in quantum probability. A striking feature of quantum probability (or noncommutative probability) is the existence of several notions of independence. This is the central topic of the remaining sections of Franz' lecture. We want to study the theory of the fundamental 'noises' in quantum probability. By a fundamental 'noise' we shall mean a quantum stochastic process with independent and stationary increments. These are generalizations of Lévy processes from classical probability, well-known special cases are Brownian motion and the Poisson process. Most other stochastic processes can be constructed from such processes, which explains why they play such a prominent role in applications.

In the fourth section of Franz' lecture we briefly review the theory of Lévy processes in classical probability. In the fifth section we treat the case of tensor independence, which is the most natural generalization to quantum probability spaces of the classical notion of independence. This notion also corresponds to the notion of independent quantum observables used by physics. We give an introduction to Schürmann's theory of Lévy processes on involutive bialgebras.

In the sixth section we develop this theory further under the assumption that the involutive bialgebra belongs to a compact quantum group in the sense of Woronowicz. These bialgebras have a richer structure and it is interesting to study Lévy processes that satisfy natural compatibility conditions with respect to this structure.

In the seventh section we introduce several other notions of independence that can be used in quantum probability. These are freeness, boolean independence, and monotone independence. For all these independences we can define convolutions. We show how these convolutions can be computed for probability measure on the real line or the unit circle using their Cauchy-Stieltjes transforms.

Next we present a classification by Muraki that states that these are the only 'nice' or 'universal' notions of independence. More precisely, they are the only possible notions on the category of noncommutative algebraic probability spaces that are based on an associative functorial product, as we shall see in the eighth section of Franz' lecture.

Finally, in the last section of Franz' lecture we study Lévy processes whose increments are independent in the sense of these notions. They are defined on dual groups, a kind of algebras similar to bialgebras, but with the usual tensor product of associative algebras replaced by their free product.

Dynamical systems. Skalski's lecture presents several examples of how the theory of dynamical systems can be generalized to a noncommutative theory. The main focus is put on two building blocks of modern abstract theory of dynamical systems: entropy and ergodic theorems. The first section of the lecture introduces the concepts and examples of quantum topological dynamical systems (understood as endomorphisms of C^*-algebras) and develops systematically the analogies with classical topological spaces and classical dynamics. The three following sections concern the most successful generalization of the notion of topological entropy to quantum dynamical systems: Voiculescu entropy. Again several examples are given, including the computation of Voiculescu entropy of the shift endomorphism on Cuntz algebras. Here also some permanence properties of Voiculescu entropy are established and the role of classical subsystems of a given quantum dynamical system is discussed. Finally in sections 5 and 6 attention is turned toward quantum "measurable" dynamical systems, understood as normal endomorphisms of von Neumann algebras. Both mean and individual ergodic theorems are treated in that context, with the latter requiring a particularly novel (in comparison with the classical set-up) approach: the notion of almost everywhere convergence, which a priori requires a concept of 'points' in the space under investigation, is replaced by the almost uniform convergence, motivated by Egorov theorem.

Throughout both lectures the balance between two crucial aspects of noncommutative mathematics is underlined: on one hand the desire to build an extension of the classical theory often necessitates providing new, original ways to obtain and understand well-known "commutative" results; on the other hand the richness of the quantum mathematical world presents completely novel phenomena, never encountered in the classical setting. We hope this interplay will enchant our readers in the way similar to how it never ceases to amaze ourselves. If QP were always equal to PQ, the world would be infinitely more boring!

1

Independence and Lévy Processes in Quantum Probability

1.1 Introduction

Quantum probability is a generalization of both classical probability theory and quantum mechanics that allows to describe the probabilistic aspects of quantum mechanics. This generalization is formulated in two steps. First, the theory is reformulated in terms of algebras of functions on probability spaces. Therefore, the notion of a probability space $(\Omega, \mathcal{F}, P)$ is replaced by the pair $(L^\infty(\Omega), E(\cdot) = \int_\Omega \cdot \,\mathrm{d}P)$ consisting of the commutative von Neumann algebra of bounded random variables and the expectation functional. Then, the commutativity condition is dropped. In this way we arrive at the notion of a (von Neumann) algebraic probability space (N, Φ) consisting of a von Neumann algebra N and a normal (faithful tracial) state Φ. As we have seen this includes classical probability spaces in the form $(L^\infty(\Omega), E)$, it also includes quantum mechanical systems modeled by a Hilbert space H and a pure state $\psi \in H$ (or a mixed state $\rho \in S(H)$), if we take $N = B(H)$ and Φ the state defined by $\Phi(X) = \langle \psi, X\psi \rangle$ (or $\Phi(X) = \mathrm{tr}(\rho X)$) for $X \in B(H)$. Note that in this course we shall relax the conditions on N and Φ and work with involutive algebras and positive normalized functionals, that is, $*$-algebraic probability spaces.

A striking feature of quantum probability (also called noncommutative probability) is the existence of several notions of independence. This is the starting point of this course, which intends to give an introduction to the theory of quantum stochastic processes with independent increments.

However, before we come to these processes, we will give a general introduction to quantum probability. In Section 1.2, we recall the basic definitions of quantum probability and discuss some fundamental differences between classical probability and quantum probability. In Section 1.3 we address the question 'Why do we need Quantum Probability?' We discuss the EPR experiment and the Kochen–Specker Theorem, which show that we cannot model quantum physics with classical probability spaces because the values of observable quantities do not exist unless we specify which measurement we will carry out and which quantities we will determine. In this sense quantum physics requires a more radical description of chance. As so far all experiments have confirmed the sometimes counterintuitive predictions of quantum physics, it follows that quantum probability is necessary to describe the real world at the microscopic level.

For the rest of the course we choose to focus on models that are stationary and have certain independence properties.

In Section 1.4, we recall the basic theory of stochastic processes with independent and stationary increments in classical probability. The marginal distributions of such processes are infinitely divisible, see Definition 1.4.1, and form convolution semigroups, see Definition 1.4.2. We recall several classification results for infinitely divisible distributions and convolution semigroups.

In Section 1.5, we start with an important class of quantum Lévy processes, that is, quantum stochastic processes with independent and stationary increments, namely those defined on involutive bialgebras. Involutive bialgebras are involutive algebras equipped with an algebra homomorphism $\Delta : \mathcal{B} \to \mathcal{B} \otimes \mathcal{B}$ from the algebra into its tensor product satisfying several conditions. This map allows to compose random variables and to define a notion of increments. The notion of independence that is used for this class of quantum Lévy processes is tensor independence, which carries its name because it is based on the tensor product of functionals and algebras. It generalizes the notion of stochastic independence used in classical probability and corresponds to the notion of independent observables in quantum physics. The basic theory of these processes is owing to Schürmann, *cf.* [Sch93].

Recently a richer, more analytic theory of quantum Lévy processes, defined on Woronowicz' compact quantum groups [Wor87a, Wor98], has been initiated, *cf.* [CFK14]. While usually defined in the C*-algebraic setting (see Definition 1.6.1), these quantum groups can also be viewed as a special class of involutive bialgebras, sometimes also called CQG algebras. CQG algebras are involutive Hopf algebras canonically associated to Woronowicz' compact quantum groups, *cf.* [Wor87a, Wor98]. They have a richer structure, in particular, an antipode and a Haar state, which satisfies a KMS property. This allows to formulate properties of a Lévy process, which guarantee that the Markov semigroup can be extended to a C*- and a von Neumann algebra, and to the associated noncommutative L_p spaces. Cipriani, Franz, and Kula have used this additional structure to apply the theory of noncommutative Markov processes and noncommutative Dirichlet forms to Lévy processes on CQG algebras, see [CFK14]. In Section 1.6 we give an introduction to compact quantum groups and show that Lévy processes on compact quantum groups are in one-to-one correspondence with time- and space-homogeneous Markov semigroups, see Theorem 1.6.6.

In noncommutative probability there exist new, truly noncommutative notions of independence that have no counterpart in classical probability. Schürmann [Sch95b] has shown that it is possible to define quantum Lévy processes with increments that are independent in the sense of these new notions of independence, if one replaces the tensor product in the theory of bialgebras by the free product of algebras. In the last three sections of this course we will give an introduction to the quantum Lévy processes obtained in this way.

In Section 1.7, we give a first introduction to these so-called universal notions of independence. We define free, monotone, and boolean independence for subalgebras of a quantum probability space and study the convolutions associated to these independences for probability measures on the real line, the positive half-line, and the unit circle.

The universal independences are based on associative universal products of algebraic probability spaces. In Section 1.8, we study the independences from the point of view of products of algebraic probability spaces and review their classifications. The results in this section are mostly owing to Ben Ghorbal and Schürmann [BGS02], and to Muraki [Mur02, Mur03].

In Section 1.9, we introduce quantum Lévy processes for universal independences. For this purpose we also introduce dual semigroups and dual groups, which are the counterparts of bialgebras and Hopf algebras. They can be obtained from bialgebras and Hopf algebras if one replaces in their definitions the tensor product by the free product of algebras, see also [Voi87, Zha91].

In Section 1.10, we close this course with a list of interesting research topics and open questions.

1.2 What is Quantum Probability?

Let us start with the most fundamental definition in quantum (or noncommutative) probability.

Definition 1.2.1 A *quantum probability space* is a pair (A, φ) consisting of a von Neumann algebra A and a normal state $\varphi : A \to \mathbb{C}$.

Remark 1.2.2 The conditions on the pair (A, φ) can be varied, depending on the applications we have in mind. In the main part of these lecture notes we will work with **-algebraic probability spaces*, which are pairs (A, φ) consisting of a unital $*$-algebra A and a normalized positive functional $\varphi : A \to \mathbb{C}$.

For a definition of a von Neumann algebra and normal functionals, and some motivation for their appearance in this context refer to Section 2.5.2 of the lecture of Adam Skalski in this volume.

Before we try to motivate the definition of a quantum probability space, let us recall the definition of a probability space given in classical probability theory.

Definition 1.2.3 A *'classical' probability space* is a triple $(\Omega, \mathcal{F}, P)$, where

- Ω is a set, the *sample space*, the set of all possible outcomes.
- $\mathcal{F} \subseteq \mathcal{P}(\Omega)$ is a σ-algebra, the set of *events*.
- $P : \mathcal{F} \to [0, 1]$ is a probability measure, it assigns to each event its *probability*.

This description of randomness is based on the idea that randomness is because of a lack of information. If we know which $\omega \in \Omega$ is realized, then there is no randomness, and we know which outcome is realized for all possible experiments. However, in general this is not the case and therefore we want to work with all possible outcomes, and random variables which that are functions of these possible outcomes.

The following example shows that — in a certain sense — quantum probability contains classical probability as a special case.

Example 1.2.4 (Classical $\subseteq$ Quantum) To a classical probability space $(\Omega, \mathcal{F}, P)$ we can associate a quantum probability space (A, φ), take

- $A = L^\infty(\Omega, \mathcal{F}, P)$, the algebra of bounded measurable functions $f : \Omega \to \mathbb{C}$, called the algebra of *random variables* or *observables*.
- $\varphi : A \ni f \mapsto E(f) = \int_\Omega f \mathrm{d}P$, which assigns to each random variable/observable its expected value.

Then A is commutative and $(\Omega, \mathcal{F}, P)$ and (A, φ) are essentially equivalent (by the spectral theorem).

However, quantum probability is more general than classical probability. This additional generality is necessary to treat classical probability theory and the probabilistic structure of quantum mechanics in a common theory.

The following example is motivated by quantum mechanics.

Example 1.2.5 (Quantum mechanics) Let H be a Hilbert space, with a unit vector $\psi \in H$ (or a density matrix $\rho \in B(H)$). Then the quantum probability space associated to (H, ψ) (or (H, ρ)) is given by

- $A = B(H)$, the algebra bounded linear operators $X : H \to H$. Self-adjoint (or normal) operators can be considered as *quantum random variables* or *observables*.
- $\varphi : B(H) \ni X \mapsto \varphi(X) = \langle \psi, X\psi \rangle$, where $\psi \in H$ is a unit vector, or, more generally, $\varphi(X) = \mathrm{tr}(\rho X)$, where ρ is a density matrix.

Note that in this book inner products are always linear on the *right* side.

States of the form $\varphi(X) = \langle \psi, X\psi \rangle$ are called *pure states* or *vector states*. Note that unit vectors $\psi \in H$ and $\psi' = e^{i\phi}\psi$ that differ only by a phase and the orthogonal projection $P_\psi = |\psi\rangle\langle\psi| : u \mapsto \langle \psi, u \rangle \psi$ onto the subspace $\mathbb{C}\psi$ spanned by those vectors all define the same state. States of the form $\varphi(X) = \mathrm{tr}(\rho X)$ are called *mixed states*, if ρ is not a rank-one projection.

Is 'quantum randomness' different from 'classical randomness'? To discuss this question let us briefly recall how the quantum probability space presented above is used to model experiments in quantum mechanics.

For simplicity let us suppose that H is a finite dimensional complex Hilbert space.

Theorem 1.2.6 *(Spectral theorem) If $X \in B(H)$ is an observable (that is, a self-adjoint operator = hermitian matrix), then it can be written as*

$$X = \sum_{\lambda \in \sigma(X)} \lambda E_\lambda$$

where $\sigma(X)$ denotes the spectrum of X (= set of eigenvalues) and E_λ the orthogonal projection onto the eigenspace of X associated to the eigenvalue λ.

Physicists associate to the observables of a quantum mechanical system, like the position or momentum of a particle, the spin of an electron, or the polarization of a photon, a self-adjoint operator on some Hilbert space H. The state of the quantum mechanical system is described by a state on the algebra $B(H)$. This state is often given in the form of a density matrix, that is, a positive operator $\rho \in B(H)$ with trace equal to one. The special case where ρ is the orthogonal projection $\rho = |\psi\rangle\langle\psi|$ onto a unit vector $\psi \in H$ corresponds to a pure state and we call ψ its state vector. Note that we will freely switch between the various mathematical descriptions, that is, state vectors, density matrices, and states (in the sense of unital positive linear functionals) for the state of a quantum system.

Von Neumann's 'Collapse' Postulate:

A measurement of an observable X with spectral decomposition

$$X = \sum_{\lambda \in \sigma(X)} \lambda E_\lambda$$

on a quantum system in the state ρ can only yield values that belong to the spectrum of X. A value $\lambda \in \sigma(X)$ occurs with probability

$$p_\lambda = \mathrm{tr}(\rho E_\lambda)$$

where tr denotes the trace. If the observed value is λ, then the state 'collapses' to

$$\tilde{\rho}_\lambda = \frac{E_\lambda \rho E_\lambda}{\mathrm{tr}(\rho E_\lambda)},$$

that is, the state of the quantum system after the measurement is described by the density matrix $\tilde{\rho}_\lambda$.

As each $\lambda \in \sigma(X)$ occurs with probability

$$\rho_\lambda = P_\rho(X = \lambda) = \mathrm{tr}(\rho E_\lambda),$$

we get

$$E_\rho(X) = \sum_{\lambda \in \sigma(X)} \lambda p_\lambda = \mathrm{tr}(\rho X)$$

for the expectation and

$$\mathrm{Var}_\rho(X) = E_\rho\big((X - E_\rho(X))^2\big) = \mathrm{tr}(\rho X^2) - \big(\mathrm{tr}(\rho X)\big)^2$$

for the variance of the observable X in the state ρ.

The simplest experiments in quantum mechanics can produce only two possible outcomes, like the measurement of the spin of a fermion (like, for example the electron) in a fixed direction in a Stern-Gerlach-type experiment, or sending a single photon through a polarization filter. Such experiments are described by the two-dimensional Hilbert space $\mathbb{C}^2$.

Example 1.2.7 (Spin of a spin-$\frac{1}{2}$ particle or polarization of a photon) Consider $H = \mathbb{C}^2$. As vectors that differ only by a phase define the same state, we can assume that the first component of a state vector in $\mathbb{C}^2$ is not negative. Therefore, the most general state vector is of the form

$$u(\theta, \phi) = \cos\frac{\theta}{2}|0\rangle + e^{i\phi}\sin\frac{\theta}{2}|1\rangle = \begin{pmatrix} \cos\frac{\theta}{2} \\ e^{i\phi}\sin\frac{\theta}{2} \end{pmatrix}$$

with $\theta \in [0,\pi]$, $\phi \in [0,2\pi)$, and $|0\rangle = |\uparrow\rangle$, $|1\rangle = |\downarrow\rangle$ form an orthonormal basis for $\mathbb{C}^2$ (for example, corresponding to the states 'spin up' and 'spin down').

Note that we used here the bra-ket notation, which is standard in quantum mechanics and also frequently used in related fields such as quantum probability and quantum information. Hilbert space vectors are denoted by so-called 'ket's' $|\text{label}\rangle$, linear functionals on the Hilbert space by 'bra's' $\langle\text{label}|$, and rank one operators by $|\text{label 1}\rangle\langle\text{label 2}|$, *cf.* [wiki bra-ket]. This notation owes its name to the 'bracket' notation $\langle\text{label 1}|\text{label 2}\rangle$ for inner products, consisting of a left part, $\langle\text{label 1}|$ called the 'bra,' and a right part, $|\text{label 2}\rangle$, called the 'ket.'

We will use this notation to name functionals and rank one operators built from vectors. That is, if $u, v \in H$, then $\langle u|$ denotes the linear functional $\langle u| : H \ni x \mapsto \langle u, x\rangle \in \mathbb{C}$ and $|u\rangle\langle v|$ the operator $|u\rangle\langle v| : H \ni x \mapsto \langle v, x\rangle u \in H$.

The vector $u(\theta,\phi)$ can be visualized as the point (θ,ϕ) on the unit sphere (*Bloch sphere*) in $\mathbb{R}^3$, that is, the vector

$$\begin{pmatrix} \cos\phi\sin\theta \\ \sin\phi\sin\theta \\ \cos\theta \end{pmatrix}.$$

Density matrices are of the form

$$\rho(x,y,z) = \frac{I + x\sigma_x + y\sigma_y + z\sigma_z}{2}$$

with $x, y, z \in \mathbb{R}$, $x^2 + y^2 + z^2 \leq 1$, where

$$I = \begin{pmatrix} 1 & 0 \\ 0 & 1 \end{pmatrix}, \sigma_x = \begin{pmatrix} 0 & 1 \\ 1 & 0 \end{pmatrix}, \sigma_y = \begin{pmatrix} 0 & -i \\ i & 0 \end{pmatrix}, \sigma_z = \begin{pmatrix} 1 & 0 \\ 0 & -1 \end{pmatrix},$$

are the Pauli matrices.

Note that the density matrix associated with to a vector state $u(\theta,\phi)$ is simply

$$|u(\theta,\phi)\rangle\langle u(\theta,\phi)| = \frac{1}{2}\begin{pmatrix} 1+\cos\theta & e^{-i\phi}\sin\theta \\ e^{i\phi}\sin\theta & 1-\cos\theta \end{pmatrix} = \rho\begin{pmatrix} \cos\phi\sin\theta \\ \sin\phi\sin\theta \\ \cos\theta \end{pmatrix}.$$

The state $\varphi_\rho : B(H) \to \mathbb{C}^2$ associated with to a density matrix ρ is the linear functional defined by

$$\varphi_\rho(X) = \mathrm{tr}(\rho X)$$

for $X \in B(H)$. If $\rho = |\psi\rangle\langle\psi|$ is pure state, that is, of the form $\rho = |\psi\rangle\langle\psi|$ for some vector $|\psi\rangle$, then this becomes

$$\varphi_{|\psi\rangle\langle\psi|} = \langle\psi, X\psi\rangle.$$

Observables (self-adjoint operators) are of the form

$$X = a|u\rangle\langle u| + b|u_\perp\rangle\langle u_\perp|,$$

with $a, b \in \mathbb{R}$, u a unit vector, $u_\perp$ orthogonal to u (unique up to a phase). In any experiment, X takes values a and b, with probabilities

$$P(X = a) = \varphi\big(|u\rangle\langle u|\big) \text{ and } P(X = b) = \varphi\big(|u_\perp\rangle\langle u_\perp|\big),$$

if φ is the state of the quantum system before the measurement.

After the experiment the state will be $|u\rangle\langle u|$, if the value a was observed, and $|u_\perp\rangle\langle u_\perp|$, if the value b was observed.

Suppose that

$$u(\theta, \phi) = \begin{pmatrix} \cos\frac{\theta}{2} \\ e^{i\phi}\sin\frac{\theta}{2} \end{pmatrix}$$

is the state vector labeled by the point

$$\begin{pmatrix} \cos\phi\sin\theta \\ \sin\phi\sin\theta \\ \cos\theta \end{pmatrix}$$

on the Bloch sphere, with $\theta \in [0, \pi]$, $\phi \in [0, 2\pi)$. Then we can take

$$u_\perp(\theta, \phi) = \begin{pmatrix} \sin\frac{\theta}{2} \\ -e^{i\phi}\cos\frac{\theta}{2} \end{pmatrix} = u(\pi - \theta, \phi + \pi)$$

for the vector orthogonal to $u(\theta, \phi)$. Note that $u_\perp(\theta, \phi)$ corresponds to the opposite point on the Bloch sphere.

Let us define

$$S(\theta,\phi) = |u\rangle\langle u| - |u_\perp\rangle\langle u_\perp|$$
$$= \begin{pmatrix} \cos\theta & e^{-i\phi}\sin\theta \\ e^{i\phi}\sin\theta & -\cos\theta \end{pmatrix}$$
$$= x\sigma_x + y\sigma_y + z\sigma_z$$

where

$$\begin{pmatrix} x \\ y \\ z \end{pmatrix} = \begin{pmatrix} \cos\phi\sin\theta \\ \sin\phi\sin\theta \\ \cos\theta \end{pmatrix}.$$

is a point on the unit sphere in $\mathbb{R}^3$.

If we measure the observable $X = S(\theta,\phi)$ on a particle whose state is given by the state vector $u(\theta',\phi')$, we get

$$\begin{aligned} P(X=+1) &= |\langle u(\theta,\phi), u(\theta',\phi')\rangle|^2 = \frac{1+\cos\vartheta}{2} \\ P(X=-1) &= \frac{1-\cos\vartheta}{2} \end{aligned} \tag{1.2.1}$$

where ϑ is the angle between the points on the Bloch sphere that correspond to $u(\theta,\phi)$ and $u(\theta',\phi')$.

We can interprete the observable $S(\theta,\phi)$ as the measurement of the spin of an electron in the direction determined by θ and ϕ. The only two possible outcomes of this experiment are '+1' and '−1', which means that the spin points in the direction of the vector

$$\begin{pmatrix} x \\ y \\ z \end{pmatrix} = \begin{pmatrix} \cos\phi\sin\theta \\ \sin\phi\sin\theta \\ \cos\theta \end{pmatrix}$$

or that it points in the opposite direction, respectively.

To each observable X in a quantum probability space we can associate a classical probability space, with $\Omega = \sigma(X)$ and $P(\{\lambda\}) = \mathrm{tr}(\rho E_\lambda)$, if $X = \sum_{\lambda\in\sigma(X)} \lambda E_\lambda$ is the spectral decomposition of X.

1.2.1 Distinguishing features of classical and quantum probability

There are several fundamental differences between a classical (or 'commutative') probability space $(\Omega, \mathcal{F}, P)$ and a quantum mechanical (or 'noncommutative') probability space such as $(B(H), \langle\psi, \cdot\psi\rangle)$.

An 'event' is the most elementary question that we can ask in an experiment, that is a random variable or observable that can take only two values (denoted by 1 and 0, or by 'true' and 'false'). In a classical probability space events are described by (measurable) subsets $E \subseteq \Omega$, which can be identified with their indicator functions $\mathbf{1}_E$, whereas in quantum probability spaces they are given by closed subspaces, which are identified with their orthogonal projections. In classical probability events form a Boolean lattice, but in quantum probability in general they form a non-Boolean (or non-distributive) lattice. This has consequences for the logical structure of quantum mechanics that can be surprising.

Recall that $E \vee F$ denotes the *join* or the smallest upper bound of E and F, that is, if E and F are events in a classical probability space, that is, measurable sets, then $E \vee F$ denotes the union of E and F, and if E and F are events in a quantum probability space, that is, orthogonal projections, then $E \vee F$ is the projection on the closed subspace generated by the ranges of E and F.

The *meet* $E \wedge F$ is the greatest lower bound of E and F. In classical probability this is the intersection of E and F; in quantum probability this is the projection on the intersection of the ranges of E and F.

Theorem 1.2.8 *Let E, F be projections on H such that $EF \neq FE$. Then*

$$E \vee F \not\leq E + F.$$

From this theorem we can derive the existence of states such that the join of two non-commuting projections has a probability that exceeds the sum of the probabilities of the two projections.

Corollary 1.2.9 *Let E, F be projections on H s.t. $EF \neq FE$. Then, for some state ρ,*

$$\mathrm{tr}\rho(E \vee F) \not\leq \mathrm{tr}\rho E + \mathrm{tr}\rho F.$$

Let us give a simple example in the setting of Example 1.2.7.

Example 1.2.10 Let $H = \mathbb{C}^2$, $\psi = |0\rangle$, $u = \cos\frac{\theta}{2}|0\rangle + \sin\frac{\theta}{2}|1\rangle$, $v = \cos\frac{\phi}{2}|0\rangle + \sin\frac{\phi}{2}|1\rangle$, with $\theta, \phi \in [0, \pi)$, and denote by E and F the projections onto $\mathbb{C}u$ and $\mathbb{C}v$, resp. If $\phi \neq \theta$, then the join of E and F is the identity and therefore $\mathrm{tr}\rho(E \vee F) = 1$. For E and F we have

$$\mathrm{tr}(\rho E) = \cos\frac{\theta}{2} \quad \text{and} \quad \mathrm{tr}(\rho F) = \cos\frac{\phi}{2}.$$

Choosing both θ and ϕ close to π, we can make these probabilities arbitrarily small.

Another consequence of the non-commutativity in quantum probability is the existence of lower bounds on the product of the variances of non-commuting observables.

Theorem 1.2.11 *(Heisenberg Uncertainty Relation) X, Y observables, ρ a state. Then*

$$\begin{aligned} \mathrm{Var}_\rho(X)\mathrm{Var}_\rho(Y) &\geq \left(\mathrm{tr}\rho\frac{1}{2}(XY + YX)\right)^2 + \left(\mathrm{tr}\rho\frac{i}{2}(XY - YX)\right)^2 \\ &\geq \frac{1}{2}\left(\mathrm{tr}\rho i[X, Y]\right)^2. \end{aligned}$$

This theorem can be demonstrated by applying the Kadison–Schwarz inequality (stated in Section 2.6.1 in Skalski's lecture in this volume) to $\mathrm{tr}\rho(X + zY)^*(X + zY)$ and carefully choosing the value for z, see also [Par03, Proposition 1.2.3].

The bracket in Heisenberg's Uncertainty Relation denotes the commutator, $[X, Y] = XY - YX$. If we apply this relation to the position operator Q and the moment operator P, then we have $[P, Q] = iI$, where I denotes the identity operator, and

$$\mathrm{Var}_\rho(Q)\mathrm{Var}_\rho(P) \geq \frac{1}{2}$$

that is, there exists no state in which the variances of position and momentum are both small.

Another difference concerns the structure of the set of extreme points of the set of states of a classical and a quantum probability space, see also the dictionary 'classical $\leftrightarrow$ quantum" below.

Theorem 1.2.12 *(Extreme points) The set of extreme states (that is, pure states) on an n-dimensional complex Hilbert space is a real manifold of dimension $2n - 2$.*

The set of all extreme probability measures on a sample space of n points has cardinality n (the probability measures δ_ω that are concentrated in one point).

1.2.2 Dictionary 'Classical $\leftrightarrow$ Quantum'

We summarize the basic terminology in classical and quantum probability, following Parthasarathy, *cf.* [Par03]. For simplicity, we shall restrict ourselves to 'finite' probability spaces, that is, assume that Ω is a finite set and H has finite dimension.

	Classical	Quantum
Sample space	A set $\Omega = \{\omega_1, \ldots, \omega_n\}$	A Hilbert space $H = \mathbb{C}^n$
Events	Subsets of Ω that form a σ-algebra (also a Boolean algebra)	Orthogonal projections in H, they form a lattice, which is not Boolean (or distributive), for example, in general $E \wedge (F_1 \vee F_2) \neq (E \wedge F_1) \vee (E \wedge F_2)$
Random variables/ observables	Measurable functions $f : \Omega \to \mathbb{R}$, they form a comm. (von Neumann) algebra, to each event $E \in \mathcal{F}$ we get a r.v. $\mathbf{1}_E$.	Self-adjoint operators $X : H \to H$, $X^* = X$, they span a non-comm. (von Neumann) algebra, events are observables with values in $\{0, 1\}$. Note that $E_\lambda = \mathbf{1}_{\{\lambda\}}(X)$.
Probability distribution/ state	An additive function $P : \mathcal{F} \to [0,1]$ determined by n pos. real numbers $p_k = P(\{\omega_k\})$ s.t. $\sum_{k=1}^n p_k = 1$ $P(E) = \sum_{\omega \in E} P(\{\omega\})$	A density matrix, that is, a pos. operator with $\mathrm{tr}(\rho) = 1$ $P(X = \lambda) = \mathrm{tr}(\rho E_\lambda)$, $P(X \in E) = \mathrm{tr}(\rho \mathbf{1}_E(X))$, $\mathbf{1}_E(X) = \sum_{\lambda \in E \cap \sigma(X)} E_\lambda$.
Expectation	$E(f) = \int_\Omega f \mathrm{d}P = \sum_{k=1}^n f(\omega) P(\{\omega\})$	$E(X) = \mathrm{tr}(\rho X)$
Variance	$\mathrm{Var}(f) = E(f^2) - E(f)^2$	$\mathrm{Var}(X) = E(X^2) - E(X)^2 = \mathrm{tr}(\rho X^2) - (\mathrm{tr}(\rho X))^2$.

cont...

continued

Extreme points	The set of all probab. distributions on Ω is a compact convex set with n extreme points $\delta_{\omega_k}, k = 1, \ldots, n$. If $P = \delta_{\omega_k}$, then the distribution of any r.v. is concentrated in one point.	The extreme points of the set $\mathcal{S}(H)$ of states on H are exactly the one-dim. projections onto the rays $\mathbb{C}u$, $u \in H$ a unit vector. If $\rho = P_u$ then $\text{Var}(X) =$ $= \|(X - \langle u, Xu \rangle)u\|^2$, thus $\text{Var}(X) = 0$ iff u is an eigenvector of X.

In quantum probability or quantum physics, *degeneracy of the state does not kill the uncertainty of the observables*. The gives an indication that the nature or origin of the randomness in quantum mechanics is fundamentally different from the generally accepted description in classical probability. One might call extreme states, that is, density matrices that are rank one operators, *degenerate*, by analogy with the situation in classical probability where they correspond to Dirac measures. However, in quantum probability this situation is very different from the one in classical probability. Even if we have maximal information on the state, that is, if the state is pure and therefore an extreme point of the set of states, the

	Classical	Quantum
Product spaces systems	Given two systems described by $(\Omega_i, \mathcal{F}_i, P_i)$, $i = 1, 2$, $(\Omega_1 \times \Omega_2, \mathcal{F}_1 \otimes \mathcal{F}_2, P_1 \otimes P_2)$ describes both independent systems as a single system	Given two systems described by (H_i, ρ_i), $i = 1, 2$, $(H_1 \otimes H_2, \rho_1 \otimes \rho_2)$ describes both indep. systems as a single system $\rightarrow$ independence $\rightarrow$ entanglement
Reversible dynamics	is modeled by bijective (measurable) maps $T : \Omega \rightarrow \Omega$ for r.v.: $f \mapsto f \circ T$ for prob. measures $P \mapsto P \circ T^{-1}$	is modeled by unitary operators $U : H \rightarrow H$ in the Heisenberg picture: $X \mapsto U^* X U$ in the Schrödinger $\rho \mapsto U \rho U^*$ or $\psi \mapsto U\psi$.

randomness does not disappear, except for some special observables (those for which the state vector is an eigenvector).

Entanglement of states of a composed system is a new phenomenon in quantum physics that is behind many of the (apparent) paradoxes. It is also used as a kind of resource in quantum computation and quantum communications. For more information, see [Par03] and the references therein.

Not necessarily reversible dynamics can more generally be described by completely positive maps. See Adam Skalski's lecture 'Quantum dynamical systems from the point of view of noncommutative mathematics' in this volume.

1.3 Why do we Need Quantum Probability?

Owing to the spectral theorem the descriptions of an individual observable as an element in a quantum probability space or by random variable defined on a classical probability space are equivalent, see also Example 1.2.4. This extends to commuting families of observables, but not to non-commuting observables. In this section we shall discuss several experiments and theorems that suggest that it is indeed impossible to describe quantum mechanics 'in a reasonable way' via classical probability spaces.

1.3.1 Mermin's version of the EPR experiment

In this section we will discuss the EPR experiment, named after Einstein, Podolsky and Rosen, in a version introduced by Mermin [Mer85], see also [Pen89].

We consider a quantum system that consists of two particles that can be described by the two-dimensional Hilbert space $\mathbb{C}^2$. We keep the notation from Example 1.2.7. The joint system is therefore described by the four-dimensional Hilbert space

$$\mathbb{C}^4 \cong \mathbb{C}^2 \otimes \mathbb{C}^2 \cong \operatorname{span}\{|jk\rangle = |j\rangle \otimes |k\rangle; j,k = 0,1\}.$$

We assume that the particles are prepared in the state with state vector

$$\psi = \frac{1}{\sqrt{2}}\big(|01\rangle - |10\rangle\big).$$

If the particles are electrons, then this state is a superposition of first particle 'up' and second particle 'down', and first particle 'down' and second particle 'up', the total spin is therefore zero.

Note that ψ does not depend on the direction that we choose for our coordinate system, if we take

$$\begin{aligned} u(\theta,\phi) &= \cos\frac{\theta}{2}|0\rangle + e^{i\phi}\sin\frac{\theta}{2}|1\rangle, \\ u_\perp(\theta,\phi) &= \sin\frac{\theta}{2}|0\rangle - e^{i\phi}\cos\frac{\theta}{2}|1\rangle, \end{aligned}$$

instead of the vectors $|1\rangle$ and $0\rangle$, with $\theta \in [0,\pi]$, $\phi \in [0,2\pi)$, then we have

$$\begin{aligned} \psi(\theta,\phi) &= \frac{1}{\sqrt{2}}\big(|u(\theta,\phi)\rangle \otimes |u_\perp(\theta,\phi)\rangle - |u_\perp(\theta,\phi)\rangle \otimes |u(\theta,\phi)\rangle\big) \\ &= \frac{1}{\sqrt{2}}\Big(\cos\frac{\theta}{2}\sin\frac{\theta}{2}|00\rangle - e^{i\phi}\cos^2\frac{\theta}{2}|01\rangle + e^{i\phi}\sin^2\frac{\theta}{2}|10\rangle \\ &\quad -e^{2i\phi}\cos\frac{\theta}{2}\sin\frac{\theta}{2}|11\rangle - \cos\frac{\theta}{2}\sin\frac{\theta}{2}|00\rangle + e^{i\phi}\cos^2\frac{\theta}{2}|10\rangle \\ &\quad -e^{i\phi}\sin^2\frac{\theta}{2}|01\rangle + e^{2i\phi}\cos\frac{\theta}{2}\sin\frac{\theta}{2}|11\rangle\Big) \\ &= \frac{-e^{i\phi}}{\sqrt{2}}(|01\rangle - |10\rangle) = -e^{i\phi}\psi, \end{aligned} \tag{1.3.1}$$

that is, the vector $\psi(\theta,\phi)$ implements the same state as ψ,

We suppose that each of two physicists, called Alice and Bob, receives one of the two particles. Let

$$S_A(\theta,\phi) = S(\theta,\phi)\otimes \mathrm{id} \quad \text{and} \quad S_B(\theta,\phi) = \mathrm{id}\otimes S(\theta,\phi)$$

where $S(\theta,\phi)$ denotes the observable corresponding to a spin measurement on an electron defined in Example 1.2.7. Then $S_A(\theta,\phi)$ corresponds to a spin measurement on Alice' particle, and $S_B(\theta,\phi)$ corresponds to a spin measurement on Bob's particle. The spectral decompositions of these observables are given by

$$\begin{aligned} S_A(\theta,\phi) &= |u\rangle\langle u|\otimes \mathrm{id} - |u_\perp\rangle\langle u_\perp|\otimes \mathrm{id}, \\ S_B(\theta,\phi) &= \mathrm{id}\otimes |u\rangle\langle u| - \mathrm{id}\otimes |u_\perp\rangle\langle u_\perp|, \end{aligned}$$

and it is straight forward to check that we have

$$P\big(S_A(\theta,\phi) = \pm 1\big) = \frac{1}{2} = P\big(S_B(\theta,\phi) = \pm 1\big)$$

for any $0 \le \theta \le \pi$, $0 \le \phi < 2\pi$. This means that Alice and Bob will observe that the two results '+1' and '−1', or 'up' and 'down', are equally likely, no matter what direction they choose for their experiment. However, after Alice has observed the spin in some direction, say $S_A(\theta,\phi)$, then the state has to be updated according to the result of her experiment. If Alice observes '+1', that is, 'up' in the direction of her choice, say, then, according to von Neumann's 'Collapse' postulate, the state after the experiment is

$$\frac{\big(|u(\theta,\phi)\rangle\langle u(\theta,\phi)|\otimes \mathrm{id}\big)\psi}{\big|\big|\big(|u(\theta,\phi)\rangle\langle u(\theta,\phi)|\otimes \mathrm{id}\big)\psi\big|\big|} = u_\perp(\theta,\phi)\otimes u(\theta,\phi),$$

and if she observes '−1', that is, 'down,' then the new state is

$$\frac{\big(|u_\perp(\theta,\phi)\rangle\langle u_\perp(\theta,\phi)|\otimes \mathrm{id}\big)\psi}{\big|\big|\big(|u_\perp(\theta,\phi)\rangle\langle u_\perp(\theta,\phi)|\otimes \mathrm{id}\big)\psi\big|\big|} = u(\theta,\phi)\otimes u_\perp(\theta,\phi)$$

(where we neglected the phase, since it does not change the state). This calculation becomes particularly easy, if we use Equation (1.3.1) to write ψ in terms of $u(\theta,\phi)$ and $u_\perp(\theta,\phi)$.

Therefore, if Bob now measures the spin of his particle in some direction, say $S_B(\theta',\phi')$, then he has to use the state $u_\perp(\theta,\phi)$ or $u(\theta,\phi)$, and his probability to observe '+1' or '−1' depends on the angle between the vectors

$$\begin{pmatrix} -\cos\phi\sin\theta \\ -\sin\phi\sin\theta \\ -\cos\theta \end{pmatrix} \quad \text{or} \quad \begin{pmatrix} \cos\phi\sin\theta \\ \sin\phi\sin\theta \\ \cos\theta \end{pmatrix}, \text{ resp.}$$

and

$$\begin{pmatrix} \cos\phi'\sin\theta' \\ \sin\phi'\sin\theta' \\ \cos\theta' \end{pmatrix},$$

according to Equation (1.2.1).

We pretended here that Alice's measurement takes place before Bob's; however, in fact the joint probabilities of Alice and Bob's measurements do not depend on the order of their measurements. Their observables commute and are therefore independent; therefore, the results are the same if they measure simultaneously or one after the other.

If they measure in directions that are at an angle ϑ, then joint probabilities of Alice and Bob's results are

	'+1' for Bob	'−1' for Bob
'+1' for Alice	$\frac{1-\cos\vartheta}{4}$	$\frac{1+\cos\vartheta}{4}$
'−1' for Alice	$\frac{1+\cos\vartheta}{4}$	$\frac{1-\cos\vartheta}{4}$

In particular, if they measure the spin in the same direction, they always find opposite results.

In Mermin's version of the EPR experiment, Alice and Bob receive a large number of particles prepared in the state ψ. They agree on three directions, say a, b, c, that are at an angle of 120° to one another, randomly measure the spin in one of these directions, record their results, and then compare them.

When they measure in the same directions, one of them will find '+1' and the other will find '−1', by the nature of the state ψ.

Suppose now that they measure in different directions, say a and b. If Alice measures in direction a and finds '+1', then the particle on which Bob carries out his measurement arrives in a state pointing in the opposite direction of a, which has an angle of 60° with the direction he has chosen. Therefore, he will find '+1' with probability $\frac{3}{4} = \frac{1+\cos 60^\circ}{2}$, and '−1' with probability $\frac{1}{4} = \frac{1-\cos 60^\circ}{2}$. Had Alice measured '−1', then Bob's particle will point in the direction of a, which has an angle of 120° with the direction he has chosen. Therefore he will find '+1' with probability $\frac{1}{4} = \frac{1+\cos 120^\circ}{2}$, and '−1' with probability $\frac{3}{4} = \frac{1-\cos 120^\circ}{2}$.

In any case, if they choose the directions independently and with a uniform probability, their results agree with probability $\frac{1}{2}$ and disagree with probability $\frac{1}{2}$, since

$$\begin{aligned} P(\text{same result}) &= P(\text{same result}|\text{same direction})\, P(\text{same direction}) \\ &\quad + P(\text{same result}|\text{diff. direction})\, P(\text{diff. direction}) \\ &= 0 \cdot \frac{1}{3} + \frac{3}{4} \cdot \frac{2}{3} \\ &= \frac{1}{2}. \end{aligned}$$

It is a remarkable fact that this behavior cannot be reproduced with classical random variables, that is, there do not exist any classical random variables

$$A_a, A_b, A_c, B_a, B_b, B_c$$

such that

$$P(A_x = -B_x) = 1$$

for all $x \in \{a, b, c\}$ and

$$P(A_{X_1} = B_{X_2}) = \frac{1}{2}$$

for X_1, X_2 two independent random variables with uniform distribution over $\{a, b, c\}$, independent of $A_a, A_b, A_c, B_a, B_b, B_c$.

Let us verify that this is indeed impossible. The first condition obviously implies $A_x = -B_x$ for $x \in \{a, b, c\}$. Therefore, we can only choose half of the random variables $A_a, A_b, A_c, B_a, B_b, B_c$, say, A_a, A_b, A_c.

If we choose all A's equal to '+1', then all B's are equal to '−1', and, no matter how Alice and Bob choose the directions of their measurements, they will always find opposite values.

If we choose

$$(A_a, A_b, A_c) = (+1, +1, -1),$$

then

$$(B_a, B_b, B_c) = (-1, -1, +1),$$

and we see that Alice and Bob get the same result only if they measure in the directions $(a, c), (b, c), (c, a), (c, b)$, that is, only for four of the nine possible combinations.

The remaining cases are similar, for any choice of values for (A_a, A_b, A_c) the probability that Alice and Bob get the same result is at most $\frac{4}{9}$, if they randomly choose their directions. Therefore, it is impossible to reproduce the quantum probabilistic joint probabilities with classical random variables. No matter how we define (A_a, A_b, A_c), the probability that Alice and Bob observe the same result when they choose their directions uniformly and independently can never exceed $\frac{4}{9}$.

Therefore, if quantum mechanics is correct, as all experiments have confirmed until now, then it is impossible to assign values to the observables, independently of the measurements we choose.

It is possible to define 18 random variables $\big(A_{(x,y)}, B_{(x,y)}\big)_{(x,y)\in\{a,b,c\}^2}$ that do produce the correct joint probabilities, see for example the discussion in [Gil98]. However, in such a model the random variable that describes the spin measured by Alice depends on the choice of the direction of Bob's spin measurement. As these experiments may be carried out simultaneously in two places that are far away, they should not influence each other.

1.3.2 Gleason's theorem

Gleason's theorem shows that for Hilbert spaces with $\dim H \geq 3$ states on $B(H)$ are in one-to-one correspondence with probability measures on $\mathcal{P}(H)$ (to be defined below). This is not the case for $\dim H = 2$; there exist many additive probability measures on $\mathcal{P}(\mathbb{C}^2)$ that do not extend to states on $B(\mathbb{C}^2)$.

Let H be a separable Hilbert space and denoted by $\mathcal{P}(H)$ the set of orthogonal projections on H.

Definition 1.3.1 A map $P : \mathcal{P}(H) \to [0,1]$ is called an *additive probability measure* on $\mathcal{P}(H)$ if $P(\mathrm{id}_H) = 1$ and

$$P\left(\sum_{k=1}^{n} E_k\right) = \sum_{k=1}^{n} P(E_k)$$

for all $n \geq 1$ and all families $E_1, \ldots, E_n$ of pairwise orthogonal projections.

If the additivity condition holds also for countable families, then we say that P is a *σ-additive probability measure* on $\mathcal{P}(H)$.

Gleason's theorem says that for a Hilbert space of dimension bigger than or equal to three all probability measures on $\mathcal{P}(H)$ arise by restriction from states on $B(H)$.

Theorem 1.3.2 *(Gleason's theorem, [nlab Gl], [Par92, Corollary 8.10], [Ara09, Theorem 2.3]) Let* $\dim H \geq 3$. *Then each additive measure on $\mathcal{P}(H)$ can be uniquely extended to a state on $B(H)$. Conversely, the restriction of every state to $\mathcal{P}(H)$ is an additive measure on $\mathcal{P}(H)$.*

The same holds for σ-additive probability measures and normal states: Every σ-additive probability measure can be extended to a normal state and every normal state restricts to a σ-additive probability measure.

For a proof of Gleason's theorem see, for example, [Par92].

1.3.3 The Kochen–Specker theorem

We will now review the Kochen–Specker theorem, which again states that it is impossible to assign values to quantum mechanical observables independently of a choice of measurements.

Definition 1.3.3 A *valuation* (or 'dispersion-free' additive probability measure) is an additive probability measure $\mu : \mathcal{P}(H) \to \{0,1\}$.

A valuation is a function that assigns to any event either the value '1' (='true') or the value '0' (='false') in a consistent way, that is, in any collection of mutually orthogonal events there is at most one that is true. In a *realistic* model of a quantum system the outcome of any possible experiment should exist no matter which experiment we will carry out. In a *non-contextual* realistic model these outcomes should not depend on the 'context', that is, on the choice of other independent experiments. E.g., in the EPR experiment the spin measured for one particle should not depend on the choice of the direction of the spin measurement carried out on the other distant particle. In this case one speaks also of a *local* model.

Non-contextual realistic models with 'hidden variables' are similar to classical probabilistic models. If we know $\omega \in \Omega$ is realized, then we can predict the outcome of all possible experiments; however, there might be obstacles that make it impossible to know the values of the hidden variables.

The Kochen–Specker Theorem shows that it is actually impossible to assign consistently truth values to the events on a Hilbert space with dimension bigger than or equal to three, that is, that such hidden variables not only cannot be known, but that they cannot exist. In our context it says that there cannot exist any classical probability space with events that correspond to the events of a quantum mechanical probability space $(B(H), \varphi)$ with $\dim H \geq 3$.

Theorem 1.3.4 *(Kochen–Specker theorem)[nlab KS] Let* $\dim(H) \geq 3$. *Then there exists no valuation on* $\mathcal{P}(H)$.

If such a valuation $\mu : \mathcal{P}(H) \to \{0,1\}$ existed, then it would extend to a state φ on $B(H)$, by Gleason's theorem. Let us assume that either $\dim(H) < \infty$ or that μ is σ-additive, so that the state φ is normal.

In the GNS representation of this state (see the lecture 'Quantum dynamical systems from the point of view of noncommutative mathematics' by Adam Skalski in this volume for an introduction to the GNS representation), the projections on which the valuation takes the value 1 would leave the cyclic vector invariant, whereas the projections on which the valuation takes the value 0 would map the cyclic vector to zero, since

$$\|P\Omega\|^2 = \langle P\Omega, P\Omega\rangle = \langle \Omega, P\Omega\rangle = \varphi(P)$$

for any projection P. As the linear span of $\mathcal{P}(H)$ is weakly dense in $B(H)$, it would follow that the GNS representation of φ is one-dimensional, which is absurd.

The classical proofs of the Kochen–Specker Theorem proceed by finding a family of orthonormal bases of H such that it is impossible to assign to the basis vectors the values 0 and 1 in such a way that any basis contains exactly one vector with the value 1. See [KS67, GK13].

This theorem proves that there exist no 'realistic non-contextual' models of quantum probability. If quantum mechanics is correct, as all experiments carried out so far seem to indicate, then it is impossible to describe quantum systems in a realistic non-contextual way by classical probability spaces.

It is possible to describe, e.g., the EPR experiment by a *contextual* commutative probability space, *cf.* [Gil98], but in a *non-contextual* model the measurement of one observable should not depend on the choice that other independent measurements are carried out in parallel.

The standard choice in conventional quantum physics is to give up realism. The outcome of an experiment does not exist until we have chosen which experiment we will carry out. In this sense the randomness of an experiment described by a quantum probability space has a different, more fundamental flavor than the way in which classical probability spaces describe randomness. The Wikipedia page [wiki Bell] gives more information on various versions of Bell's inequalities, for example, the CHSH inequality, their practical experimental tests, their interpretation, and a few references for further reading.

1.4 Infinite Divisibility in Classical Probability

Before we study quantum stochastic processes with independent and stationary increments, that is, quantum Lévy processes, let us first recall some definitions and facts about infinite divisibility and Lévy processes in classical probability. See also [Sko91, Ber98, Sat99, App04, App05, Kyp07].

1.4.1 Stochastic independence

Recall that two random variables $X_1 : (\Omega, \mathcal{F}, P) \to (E_1, \mathcal{E}_1)$ and $X_2 : (\Omega, \mathcal{F}, P) \to (E_2, \mathcal{E}_2)$ are called *independent* if their joint law $P_{(X_1,X_2)}$ is equal to the product of their marginal laws, that is,

$$P_{(X_1,X_2)} = P_{X_1} \otimes P_{X_2}.$$

This means that

$$P(X_1 \in A_1, X_2 \in A_2) = P(X_1 \in A_1)P(X_2 \in A_2)$$

for all $A_1 \in \mathcal{E}_1$, $A_2 \in \mathcal{E}_2$.

1.4.2 Convolution

Let G be a topological semigroup with neutral element e and multiplication $m : G \times G \to G$. Then G is a measurable space with Borel-σ-algebra $\mathcal{B}(G)$. We can define the *convolution product* $\mu_1 \star \mu_2$ of two probability measures μ_1, μ_2 defined on $(G, \mathcal{B}(G))$ as the image measure $m_*(\mu_1 \otimes \mu_2)$ of their product $\mu_1 \otimes \mu_2$, that is,

$$(\mu_1 \star \mu_2)(A) = (\mu_1 \otimes \mu_2)(\{(g_1, g_2) \in G \times G; g_1 g_2 \in A\})$$

for $A \in \mathcal{B}(G)$.

If $X_1, X_2 : (\Omega, \mathcal{F}, P) \to (G, \mathcal{B}(G))$ are two independent random variables with distributions $P_{X_1} = \mu_1$, $P_{X_2} = \mu_2$, then their product has distribution

$$P_{X_1 X_2} = m_*(P_{X_1} \otimes P_{X_2}) = P_{X_1} \star P_{X_2} = \mu_1 \star \mu_2.$$

1.4.3 Infinite divisibility, continuous convolution semigroups, and Lévy processes

Definition 1.4.1 A probability measure μ on a topological semigroup G is called *infinitely divisible,* if for every integer $n \geq 1$ there exists a probability measure μ_n such that

$$\mu = \underbrace{\mu_n \star \cdots \star \mu_n}_{n \text{ times}}.$$

Definition 1.4.2 A family $(\mu_t)_{t\geq 0}$ of probability measures on a topological semigroup is called a *continuous convolution semigroup (ccs)* if

(i) $\lim_{t\searrow 0} \mu_t = \delta_e$ weakly, that is, $\lim_{t\searrow 0} \int_G f \mu_t = f(e)$ for all $f \in C_b(G)$.

(ii) $\mu_s \star \mu_t = \mu_{s+t}$ for all $s, t \geq 0$.

Definition 1.4.3 A probability measure μ is called *embeddable into a continuous convolution semigroup* if there exists a continuous convolution semigroup $(\mu_t)_{t\geq 0}$ such that $\mu = \mu_1$.

Clearly, a probability measure that is embeddable into a continuous convolution semigroup is also infinitely divisible. On many groups, e.g., $(\mathbb{R}^d, +)$ the converse is also true, but there exist also groups where the converse does not hold.

Recall that a *stochastic process* $(X_\iota)_{\iota\in I}$ is simply a family of random variables indexed by some set I.

Definition 1.4.4 A stochastic process $(X_{st})_{0\leq s\leq t}$ with values in a topological semigroup is called a (right) *Lévy process,* if

(i) (increment property) $X_{ss} = e$ and $X_{st}X_{tu} = X_{su}$ a.s. for all $0 \leq s \leq t \leq u$;

(ii) (independence) the increments $X_{s_1t_1}, \ldots, X_{s_nt_n}$ are independent for all $n \geq 1$ and all $s_1 \leq t_1 \leq s_2 \leq \cdots \leq t_n$;

(iii) (stationarity) $P_{X_{st}} = P_{X_{s+h,t+h}}$ for all $h > 0$ and all $0 \leq s \leq t$, that is, the law of X_{st} depends only on $t - s$;

(iv) (weak continuity) $(X_{st})_{0\leq s\leq t}$ is stochastically continuous, that is, we have $X_{st} \stackrel{t\searrow s}{\longrightarrow} X_{ss}$ in probability.

We define $X_t = X_{0t}$. If G is a group, then the increments can be recovered from $(X_t)_{t\geq 0}$ by $X_{st} = X_{0s}^{-1}X_{0t}$.

A stochastic process $(X_t)_{t\geq 0}$ indexed by $\mathbb{R}_+$ and with values in a group is called a Lévy process, if its increment process $(X_{st})_{0\leq s\leq t}$ with $X_{st} = X_s^{-1}X_t$ is a Lévy process in the sense of Definition 1.4.4.

Proposition 1.4.5 *If (X_{st}) is a Lévy process with values in a topological semigroup G, then its marginal distributions $\mu_t = P_{X_{0t}}$ form a continuous convolution semigroup.*

Exercise 1.4.6 Prove this Proposition.

Conversely, given a continuous convolution semigroup $(\mu_t)_{t\geq 0}$ of probability measures on a topological semigroup G, one can construct a Lévy process with values in G whose marginals are equal to the convolution semigroup $(\mu_t)_{t\geq 0}$.

1.4.4 The De Finetti–Lévy–Khintchine formula on $(\mathbb{R}_+, +)$

Let us start with a description of infinitely divisible probability measures on the semigroup $(\mathbb{R}_+, +)$.

Theorem 1.4.7 *A probability measure μ on $\mathbb{R}_+$ is infinitely divisible if and only if there exist $b \geq 0$ and ν a measure on $\mathbb{R}_+$ with $\int_0^\infty 1 \wedge x \mathrm{d}\nu(x) < \infty$ such that the Laplace transform*

$$\psi_\mu(\lambda) = \int_0^\infty e^{-\lambda x} \mathrm{d}\mu(x)$$

of μ has the form

$$\psi_\mu(\lambda) = \exp\left(\Phi(\lambda)\right)$$

for all $\lambda \geq 0$, with

$$\Phi(\lambda) = b\lambda + \int_0^\infty (1 - e^{-\lambda x}) \mathrm{d}\nu(x).$$

The pair (b, ν) is uniquely determined by μ.

Proof See [Ber98, p. 72]. □

Lévy processes with values in $(\mathbb{R}_+, +)$ have increasing trajectories, they called *subordinators*. The pair (b, ν) is called the *characteristics* or the *characteristic pair of μ*.

Corollary 1.4.8 *Every infinitely divisible probability measure on $\mathbb{R}_+$ is embeddable into a continuous convolution semigroup.*

1.4.5 Lévy–Khintchine formulae on cones

We also have the following generalization for proper closed cones in finite-dimensional vector spaces.

Recall that a non-empty subset K in a real or complex vector space is called a cone if the following two conditions are satisfied,

(i) $x_1, x_2 \in K$ implies $x_1 + x_2 \in K$,

(ii) $\lambda \geq 0$, $x \in K$ implies $\lambda x \in K$.

A cone is called proper if $K \neq \{0\}$ and $K \cap (-K) = \{0\}$, that is, K does not contain a straight line.

Theorem 1.4.9 *[Sko91] Let $K \subset \mathbb{R}^d$ be a proper closed cone and μ a probability measure on K. Then μ is infinitely divisible if and only if there exist $b \in K$ and ν a measure on K such that $\int_K 1 \wedge \|x\| \, d\nu(x) < \infty$ such that the Fourier transform of μ has the form*

$$\hat{\mu}(y) = \int_K e^{i\langle y,x\rangle} d\mu(x) = \exp\left(i\langle y, b\rangle + \int_K (e^{i\langle y,x\rangle} - 1) d\nu(x) \right)$$

for $y \in \mathbb{R}^d$.

In this case the Laplace transform is well defined on the dual cone

$$K' = \{y \in \mathbb{R}^d : \langle y, x\rangle \geq 0 \forall x \in K\}$$

and has the form

$$\psi_\mu(y) = \exp\left(-\langle y, b\rangle - \int_K (1 - e^{-\langle y,x\rangle}) d\nu(x) \right)$$

for $y \in K'$.

1.4.6 The Lévy–Khintchine formula on $(\mathbb{R}^d, +)$

Theorem 1.4.10 *[Sat99, App04] A probability measure μ on $\mathbb{R}^d$ is infinitely divisible if and only if its Fourier transform is of the form*

$$\hat{\mu}(u) = \int_{\mathbb{R}^d} e^{i\langle x,u\rangle} d\mu(x) \tag{1.4.1}$$

$$= \exp\left(i\langle b, u\rangle - \frac{1}{2}\langle u, Au\rangle + \int_{\mathbb{R}^d - \{0\}} (e^{i\langle u,y\rangle} - 1 - i\langle u, y\rangle \mathbf{1}_{\|y\|<1}) d\nu(y) \right),$$

for all $u \in \mathbb{R}^d$. The triple (b, A, ν) is uniquely determined by μ.

(b, A, ν) are called the *characteristics* or the *characteristic triple of μ*.

Corollary 1.4.11 *Any infinitely divisible probability measure on $\mathbb{R}^d$ is embeddable into a continuous convolution semigroup.*

1.4.7 The Markov semigroup of a Lévy process

Markov processes are thought of as memoryless. Roughly speaking, a stochastic processes $(X_t)_{t\geq 0}$ indexed by $\mathbb{R}_+$ is a Markov process if for any t the past and the future with respect to t are independent conditionally on X_t. More precisely, if we denote by

$$\mathcal{F}_t = \sigma(X_s; 0 \leq s \leq t)$$

the σ-algebra generated by the past of X_t, then, for $0 \leq s \leq t$, the conditional expectation $E(f(X_t)|\mathcal{F}_s)$ of a function of X_t should be a function of X_s and not depend on the values of the process strictly before s. Here we assume that the process $(X_t)_{t\geq 0}$ takes values in some measurable space $(\Omega, \mathcal{F})$ and that f is a bounded measurable function on Ω.

This allows to define transition operators $T_{s,t} : L^\infty(\Omega) \to L^\infty(\Omega)$ for $0 \leq s \leq t$ by requiring

$$(T_{s,t}f)(X_s) = E\big(f(X_t)\big|\mathcal{F}_s\big)$$

for $f \in L^\infty(\Omega)$, where $T_{s,t}f$ is determined only P_{X_s} almost everywhere. The Markov process $(X_t)_{t\geq 0}$ is called *time-homogeneous*, if the transition operators $T_{s,t}$ depend only on the difference $t - s$. In this case the operators $T_t = T_{0,t}$, $t \geq 0$, form a semigroup, called the Markov semigroup of $(X_t)_{t\geq 0}$.

Lévy processes are time-homogeneous Markov processes, they are even Feller processes, that is, their Markov semigroup maps $C_0(G)$ to itself. The independence of the increments allows for a simple description of their Markov semigroup $(T_t)_{t\geq 0}$,

$$(T_t f)(g) = E\big(f(gX_t)\big) = \int_G f(gg')\mathrm{d}P_{X_t}(g'), \qquad g \in G,$$

for $t \geq 0$, $f \in C_0(G)$.

For more information on Markov processes, Markov semigroups, and also the special case of group-valued Lévy processes, see [RW00a, RW00b].

1.4.8 Hunt's formula

Let G be a Lie group with Lie algebra $\mathfrak{g}$.

For a Lévy process in G we have its Markov semigroup $(T_t)_{t\geq 0}$ and its infinitesimal generator L. Hunt's theorem describes Lévy processes in G in terms of their generators.

Fix a basis $(X_j, 1 \leq j \leq n)$ of $\mathfrak{g}$ and define the dense subspace $C_2^L(G)$ by

$$C_2^L(G) = \left\{ f \in C_0(G); \begin{array}{c} X_i^L(f) \in C_0(G) \text{ and} \\ X_i^L X_j^L(f) \in C_0(G) \text{ for all } 1 \leq i,j \leq n \end{array} \right\},$$

where X_j^L denotes the action of X_j on appropriate functions on G as a left invariant vector field. $C_2^L(G)$ is a Banach space with respect to the norm

$$\|f\|_{2,L} = \|f\| + \sum_{i=1}^{n} \|X_i^L f\| + \sum_{j,k=1}^{n} \|X_j^L X_k^L f\|,$$

where

$$\|f\| = \sup_{g \in G} |f(g)|$$

denotes the supremum norm. The space $C_2^R(G)$ and the norms $\|\cdot\|_{2,R}$ are defined similarly. Note that the smooth functions of compact support are included in these spaces, that is, $C_c^\infty(G) \subseteq C_2^L(G) \cap C_2^R(G)$.

There exist functions $x_i \in C_c^\infty(G), 1 \leq i \leq n$ so that $(x_1, \ldots, x_n)$ are a system of canonical co-ordinates for G at e.

Theorem 1.4.12 (Hunt's theorem) *[Hun56] Let X be a Lévy process in G with infinitesimal generator L. Then*

(i) $C_2^L(G) \subseteq Dom(L)$.

(ii) For each $g \in G, f \in C_2^L(G)$,

$$\begin{aligned} Lf(g) &= \sum_i b^i X_i^L f(g) + \sum_{i,j} a^{ij} X_i^L X_j^L f(g) \\ &\quad + \int_{G\setminus\{e\}} (f(gh) - f(g) - y^i(g) X_i^L f(g)) \nu(dh), \end{aligned}$$

where $b = (b^1, \ldots b^n) \in \mathbb{R}^n, a = (a^{ij})$ *is a non-negative-definite, symmetric* $n \times n$ *real-valued matrix and* ν *is a Lévy measure on* $G\setminus\{e\}$.

Conversely, for any linear operator with such a representation there exists a Lévy process (unique up to stochastic equivalence) with generator L.

See also [App14, Theorem 5.3.3] for more details and an outline of the proof of this theorem. Michael Skeide has given a C^*-algebraic proof of Hunt's Theorem for compact Lie groups in [Ske99].

Exercise 1.4.13 Show how we can recover the Lévy-Khintchine formula for $\mathbb{R}^d$ from Hunt's Theorem.

1.5 Lévy Processes on Involutive Bialgebras

In this section we will give the definition of Lévy processes on involutive bialgebras, *cf.* Subsection 1.5.1, and develop their general theory.

In Subsection 1.5.2, we will see that the marginal distributions of a Lévy process form a convolution semigroup of states. We will associate a generating functional with a Lévy process on an involutive bialgebra, which characterizes uniquely its distribution, like in classical probability. By a GNS-type construction we can get a so-called Schürmann triple from the generating functional, see Subsection 1.5.3.

This Schürmann triple can be used to obtain a realization of the process on a symmetric Fock space. This realization can be found as the unique solution of a quantum stochastic differential equation. It establishes the one-to-one correspondence between Lévy processes, convolution semigroups of states, generating functionals, and Schürmann triples. We will not present the representation theorem here, but refer to [Sch93, Chapter 2].

Finally, in Subsection 1.5.4, we look at several examples.

For more information on Lévy processes on involutive bialgebras, see also [Sch93], [Mey95, Chapter VII], [FS99], [Fra06b].

1.5.1 Definition of Lévy processes on involutive bialgebras

In this Section and the following we will usually deal with *quantum probability spaces* in the $*$-algebraic sense, that is, pairs $(\mathcal{A}, \Phi)$ consisting of a unital $*$-algebra $\mathcal{A}$ and a state (that is, a normalized positive linear functional) Φ on $\mathcal{A}$. Positivity in this purely algebraic context simply means $\Phi(a^*a) \geq 0$ for all $a \in \mathcal{A}$. A *quantum random variable* j over a quantum probability space $(\mathcal{A}, \Phi)$ on a $*$-algebra $\mathcal{B}$ is a unit-preserving $*$-algebra homomorphism $j : \mathcal{B} \to \mathcal{A}$. A *quantum stochastic process* is an indexed family of random variables $(j_t)_{t\in I}$. For a quantum random variable $j : \mathcal{B} \to \mathcal{A}$ we will call $\varphi_j = \Phi \circ j$ its *distribution* in the state Φ. For a quantum stochastic process $(j_t)_{t\in I}$ the functionals $\varphi_t = \Phi \circ j_t : \mathcal{B} \to \mathbb{C}$ are called *marginal distributions*. The *joint distribution* $\Phi \circ (\coprod_{t\in I} j_t)$ of a quantum stochastic process is a functional on the free product $\coprod_{t\in I} \mathcal{B}$, see Section 1.8.

Two quantum stochastic processes

$$\left(j_t^{(1)} : \mathcal{B} \to \mathcal{A}_1\right)_{t\in I} \quad \text{and} \quad \left(j_t^{(2)} : \mathcal{B} \to \mathcal{A}_2\right)_{t\in I}$$

on the same $*$-algebra $\mathcal{B}$ over possibly different quantum probability spaces $(\mathcal{A}_1, \Phi_1)$ and $(\mathcal{A}_2, \Phi_2)$ are called *equivalent*, if their joint distributions coincide. This is the case, if and only if all their moments agree, that is, if

$$\Phi_1\left(j_{t_1}^{(1)}(b_1)\cdots j_{t_n}^{(1)}(b_n)\right) = \Phi_2\left(j_{t_1}^{(2)}(b_1)\cdots j_{t_n}^{(2)}(b_n)\right)$$

holds for all $n \in \mathbb{N}$, $t_1, \ldots, t_n \in I$ and all $b_1, \ldots, b_n \in \mathcal{B}$.

In Example 1.2.5 we defined quantum random variables as elements of quantum probability spaces. This can be viewed as a special case of the definition we just gave, by associating to $X \in \mathcal{A}$ the unique unital *-algebra homomorphism $j_X : \mathbb{C}\langle a, a^*\rangle \to \mathcal{A}$ such that $j_X(a) = X$, where $\mathbb{C}\langle a, a^*\rangle$ denotes the free unital algebra over $\{a, a^*\}$ with the obvious involution.

Similarly, the term 'quantum stochastic process' is sometimes also used for an indexed family $(X_t)_{t\in I}$ of operators on a Hilbert space or more generally of elements of a quantum probability space. We will reserve the name *operator process* for this. An operator process $(X_t)_{t\in I} \subseteq \mathcal{A}$ (where $\mathcal{A}$ is a $*$-algebra of operators) always defines a quantum stochastic process $(j_t : \mathbb{C}\langle a, a^*\rangle \to \mathcal{A})_{t\in I}$ on the free $*$-algebra with one generator, if we set $j_t(a) = X_t$ and extend j_t as a $*$-algebra homomorphism. On the other hand, operator processes can be obtained from quantum stochastic processes $(j_t : \mathcal{B} \to \mathcal{A})_{t\in I}$ by choosing an element x of the algebra $\mathcal{B}$ and setting $X_t = j_t(x)$.

The notion of independence we use for Lévy processes on involutive bialgebras is the so-called tensor or boson independence. In Section 1.8 we will see that other interesting notions of independence exist.

Definition 1.5.1 Let $(\mathcal{A}, \Phi)$ be a quantum probability space and $\mathcal{B}$ a $*$-algebra. The quantum random variables $j_1, \ldots, j_n : \mathcal{B} \to \mathcal{A}$ are called *tensor or Bose-independent* (w.r.t. the state Φ), if

(i) $\Phi(j_1(b_1) \cdots j_n(b_n)) = \Phi(j_1(b_1)) \cdots \Phi(j_n(b_n))$ for all $b_1, \ldots, b_n \in \mathcal{B}$, and

(ii) $[j_l(b_1), j_k(b_2)] = 0$ for all $k \neq l$ and all $b_1, b_2 \in \mathcal{B}$.

Recall that an *involutive bialgebra* $(\mathcal{B}, \Delta, \varepsilon)$ is a unital $*$-algebra $\mathcal{B}$ with two unital $*$-homomorphisms $\Delta : \mathcal{B} \to \mathcal{B} \otimes \mathcal{B}$, $\varepsilon : \mathcal{B} \to \mathbb{C}$ called *coproduct* or *comultiplication* and *counit*, satisfying

$$(\mathrm{id} \otimes \Delta) \circ \Delta = (\Delta \otimes \mathrm{id}) \circ \Delta \qquad \text{(coassociativity)}$$

$$(\mathrm{id} \otimes \varepsilon) \circ \Delta = \mathrm{id} = (\varepsilon \otimes \mathrm{id}) \circ \Delta \qquad \text{(counit property).}$$

Let $j_1, j_2 : \mathcal{B} \to \mathcal{A}$ be two linear maps with values in some algebra $\mathcal{A}$, then we define their *convolution* $j_1 \star j_2$ by

$$j_1 \star j_2 = m_{\mathcal{A}} \circ (j_1 \otimes j_2) \circ \Delta.$$

Here $m_{\mathcal{A}} : \mathcal{A} \otimes \mathcal{A} \to \mathcal{A}$ denotes the multiplication of $\mathcal{A}$, $m_{\mathcal{A}}(a \otimes b) = ab$ for $a, b \in \mathcal{A}$.

We shall frequently use *Sweedler notation* for the coproduct of an element $b \in \mathcal{B}$, that is, omit the summation and the index in the formula $\Delta(b) = \sum_i b_{(1),i} \otimes b_{(2),i}$ and write simply $\Delta(b) = b_{(1)} \otimes b_{(2)}$.

With this notation we write $(j_1 \star j_2)(b) = j_1(b_{(1)}) j_2(b_{(2)})$. If j_1 and j_2 are two independent quantum random variables, then $j_1 \star j_2$ is again a quantum random variable, that is, a $*$-homomorphism. The fact that we can compose quantum random variables allows us to define Lévy processes, that is, processes with independent and stationary increments.

Example 1.5.2 Let (G, e) be a semigroup with identity e. We call a function $f : G \to \mathbb{C}$ *representative*, if there exists a finite-dimensional representation $\pi : G \to M_n$ and vectors $u, v \in \mathbb{C}^n$ such that $f(g) = \langle v, \pi(g) u \rangle$ for all $g \in G$. These functions are sometimes also called *coefficient functions*. The algebra

$$\mathcal{R}(G) = \{f : G \to \mathbb{C} \text{ a representative function}\}$$

is an involutive bialgebra with pointwise multiplication and conjugation, and the coproduct $\Delta : \mathcal{R}(G) \to \mathcal{R}(G) \otimes \mathcal{R}(G) \cong \mathcal{R}(G \times G)$ and counit $\varepsilon : \mathcal{R}(G) \to \mathbb{C}$ defined by

$$\Delta f(g_1, g_2) = f(g_1 g_2),$$

for $g_1, g_2 \in G$ and $\varepsilon(f) = f(e)$.

Definition 1.5.3 Let $\mathcal{B}$ be an involutive bialgebra. A quantum stochastic process $(j_{st})_{0 \leq s \leq t}$ on $\mathcal{B}$ over some quantum probability space $(\mathcal{A}, \Phi)$ is called a *Lévy process,* if the following four conditions are satisfied.

(i) (Increment property) We have

$$\begin{aligned} j_{rs} \star j_{st} &= j_{rt} \quad \text{for all } 0 \leq r \leq s \leq t, \\ j_{tt} &= \varepsilon \mathbf{1} \quad \text{for all } 0 \leq t, \end{aligned}$$

that is, $j_{tt}(b) = \varepsilon(b)\mathbf{1}$ for all $b \in \mathcal{B}$, where $\mathbf{1}$ denotes the unit of $\mathcal{A}$.

(ii) (Independence of increments) The family $(j_{st})_{0 \leq s \leq t}$ is independent, that is, the quantum random variables $j_{s_1, t_1}, \ldots,$ $j_{s_n t_n}$ are tensor-independent for all $n \in \mathbb{N}$ and all $0 \leq s_1 \leq t_1 \leq s_2 \leq \cdots \leq t_n$.

(iii) (Stationarity of increments) The distribution $\varphi_{st} = \Phi \circ j_{st}$ of j_{st} depends only on the difference $t - s$.

(iv) (Weak continuity) The quantum random variables j_{st} converge to j_{ss} in distribution for $t \searrow s$, that is,

$$\lim_{t \searrow s} \Phi \circ j_{st}(b) = \Phi \circ j_{ss}(b)$$

for all $b \in \mathcal{B}$.

Recall that an *(involutive) Hopf algebra* $(\mathcal{B}, \Delta, \varepsilon, S)$ is an (involutive) bialgebra $(\mathcal{B}, \Delta, \varepsilon)$ equipped with a linear map called *antipode* $S : \mathcal{B} \to \mathcal{B}$ satisfying

$$S \star \mathrm{id} = \mathbf{1} \circ \varepsilon = \mathrm{id} \star S. \tag{1.5.1}$$

The antipode is unique, if it exists. Furthermore, it is an algebra and coalgebra anti-homomorphism, that is, it satisfies $S(ab) = S(b)S(a)$ for all $a, b \in \mathcal{B}$ and $(S \otimes S) \circ \Delta = \tau \circ \Delta \circ S$, where $\tau : \mathcal{B} \otimes \mathcal{B} \to \mathcal{B} \otimes \mathcal{B}$ is the *flip* $\tau(a \otimes b) = b \otimes a$. If $(\mathcal{B}, \Delta, \varepsilon)$ is an involutive bialgebra and $S : \mathcal{B} \to \mathcal{B}$ a linear map satisfying (1.5.1), then S satisfies also the relation

$$S \circ * \circ S \circ * = \mathrm{id}.$$

In particular, it follows that the antipode S of an involutive Hopf algebra is invertible. This is not true for Hopf algebras in general.

Exercise 1.5.4 Show that if $(k_t)_{t\geq 0}$ is any quantum stochastic process on an involutive Hopf algebra, then the quantum stochastic process defined by

$$j_{st} = m_{\mathcal{A}} \circ \big((k_s \circ S) \otimes k_t\big) \circ \Delta,$$

for $0 \leq s \leq t$, satisfies the increment property (1) in Definition 1.5.3.
A one-parameter stochastic process $(k_t)_{t\geq 0}$ on a Hopf $*$-algebra H is called a *Lévy process on H*, if its *increment process* $(j_{st})_{0\leq s\leq t}$ with $j_{st} = m_{\mathcal{A}} \circ \big((k_s \circ S) \otimes k_t\big) \circ \Delta$ is a Lévy process on H in the sense of Definition 1.5.3.

Let $(j_{st})_{0\leq s\leq t}$ be a Lévy process on some involutive bialgebra. We will denote the marginal distributions of $(j_{st})_{0\leq s\leq t}$ by $\varphi_{t-s} = \Phi \circ j_{st}$. Owing to the stationarity of the increments this is well defined.

Lemma 1.5.5 *The marginal distributions $(\varphi_t)_{t\geq 0}$ of a Lévy process on an involutive bialgebra $\mathcal{B}$ form a* continuous convolution semigroup of states *on $\mathcal{B}$, that is, they satisfy*

(i) *$\varphi_0 = \varepsilon$, $\varphi_s \star \varphi_t = \varphi_{s+t}$ for all $s,t \geq 0$, and $\lim_{t\searrow 0} \varphi_t(b) = \varepsilon(b)$ for all $b \in \mathcal{B}$, and*
(ii) *$\varphi_t(\mathbf{1}) = 1$, and $\varphi_t(b^*b) \geq 0$ for all $t \geq 0$ and all $b \in \mathcal{B}$.*

Proof $\varphi_t = \Phi \circ j_{0t}$ is clearly a state, since j_{0t} is a $*$-homomorphism and Φ a state.

From the first condition in Definition 1.5.3 we get

$$\varphi_0 = \Phi \circ j_{00} = \Phi(1)\varepsilon = \varepsilon,$$

and

$$\varphi_{s+t}(b) = \Phi\big(j_{0,s+t}(b)\big) = \Phi\left(\sum j_{0s}(b_{(1)}) j_{s,s+t}(b_{(2)})\right),$$

for $b \in \mathcal{B}$, $\Delta(b) = \sum b_{(1)} \otimes b_{(2)}$. Using the independence of increments, we can factorize this and get

$$\begin{aligned} \varphi_{s+t}(b) &= \sum \Phi\big(j_{0s}(b_{(1)})\big)\Phi\big(j_{s,s+t}(b_{(2)})\big) = \sum \varphi_s(b_{(1)})\varphi_t(b_{(2)}) \\ &= \varphi_s \otimes \varphi_t\big(\Delta(b)\big) = \varphi_s \star \varphi_t(b) \end{aligned}$$

for all $\in \mathcal{B}$.

The continuity is an immediate consequence of the last condition in Definition 1.5.3. □

Lemma 1.5.6 *The convolution semigroup of states characterizes a Lévy process on an involutive bialgebra up to equivalence.*

Proof This follows from the fact that the increment property and the independence of increments allow to express all joint moments in terms of the marginals. E.g., for $0 \leq s \leq t \leq u \leq v$ and $a, b, c \in \mathcal{B}$, the moment $\Phi\big(j_{su}(a) j_{st}(b) j_{sv}(c)\big)$ becomes

$$\begin{aligned}\Phi\big(j_{su}(a) j_{st}(b) j_{sv}(c)\big) &= \Phi\big((j_{st} \star j_{tu})(a) j_{st}(b) (j_{st} \star j_{tu} \star j_{uv})(c)\big) \\ &= \Phi\big(j_{st}(a_{(1)}) j_{tu}(a_{(2)}) j_{st}(b) j_{st}(c_{(1)}) j_{tu}(c_{(2)}) j_{uv}(c_{(3)})\big) \\ &= \Phi\big(j_{st}(a_{(1)} b c_{(1)}) j_{tu}(a_{(2)} c_{(2)}) j_{uv}(c_{(3)})\big) \\ &= \varphi_{t-s}(a_{(1)} b c_{(1)}) \varphi_{u-t}(a_{(2)} c_{(2)}) \varphi_{v-u}(c_{(3)}).\end{aligned}$$

□

It is possible to reconstruct a process $(j_{st})_{0 \leq s \leq t}$ directly from its convolution semigroup via an inductive limit, see [Sch93, Section 1.9] or [FS99, Section 4.5]. Therefore, we even have a one-to-one correspondence between equivalence classes of Lévy processes on $\mathcal{B}$ and convolution semigroups of states on $\mathcal{B}$.

There are other constructions of Lévy processes based on the Schürmann triple, see, e.g., [Sch93, Chapter 2], [LS05], and [SSV10].

1.5.2 The generating functional of a Lévy process

In this subsection we will meet two more objects that classify Lévy processes, namely their generating functionals and their triples (called Schürmann triples by P.-A. Meyer, see [Mey95, Section VII.1.6]).

We begin with a technical lemma.

Lemma 1.5.7

(a) *Let $\psi : \mathcal{C} \to \mathbb{C}$ be a linear functional on some coalgebra $\mathcal{C}$. Then the series*

$$\exp_\star \psi(b) \stackrel{\text{def}}{=} \sum_{n=0} \frac{\psi^{\star n}}{n!}(b) = \varepsilon(b) + \psi(b) + \frac{1}{2}\psi \star \psi(b) + \cdots$$

converges for all $b \in \mathcal{C}$.

(b) Let $(\varphi_t)_{t\geq 0}$ be a convolution semigroup on some coalgebra $\mathcal{C}$. Then the limit

$$L(b) = \lim_{t\searrow 0} \frac{1}{t}\big(\varphi_t(b) - \varepsilon(b)\big)$$

exists for all $b \in \mathcal{C}$. Furthermore, we have $\varphi_t = \exp_\star tL$ for all $t \geq 0$.

The proof of this lemma relies on the fundamental theorem of coalgebras [DNR01, Theorem 1.4.7], which states that for any element c of a coalgebra C there exists a finite-dimensional subcoalgebra C_c of C that contains c. This allows to reduce the study of convolution semigroups of functionals on coalgebras to the study of semigroups in finite-dimensional algebras, see [ASW88, Sch93].

Proposition 1.5.8 (Schoenberg correspondence) *Let $\mathcal{B}$ be an involutive bialgebra, $(\varphi_t)_{t\geq 0}$ a convolution semigroup of linear functionals on $\mathcal{B}$ and*

$$L = \lim_{t\searrow 0} \frac{1}{t}\big(\varphi_t - \varepsilon\big).$$

Then the following are equivalent.

(i) $(\varphi_t)_{t\geq 0}$ is a convolution semigroup of states.

(ii) $L : \mathcal{B} \to \mathbb{C}$ satisfies $L(\mathbf{1}) = 0$, and it is hermitian and conditionally positive, *i.e.,*

$$L(b^*) = \overline{L(b)}$$

for all $b \in \mathcal{B}$, and

$$L(b^*b) \geq 0$$

for all $b \in \mathcal{B}$ with $\varepsilon(b) = 0$.

Proof We prove only the (easy) direction (i)⇒(ii), the converse follows from the representation theorem, which can be found in [Sch93, Chapter 2].

The first property follows by differentiating $\varphi_t(\mathbf{1}) = 1$ w.r.t. t.

Let $b \in \mathcal{B}$, $\varepsilon(b) = 0$. If all φ_t are states, then we have $\varphi_t(b^*b) \geq 0$ for all $t \geq 0$ and therefore

$$L(b^*b) = \lim_{t \searrow 0} \frac{1}{t}\left(\varphi_t(b^*b) - \varepsilon(b^*b)\right) = \lim_{t \searrow 0} \frac{\varphi_t(b^*b)}{t} \geq 0.$$

Similarly, L is hermitian, since all φ_t are hermitian. □

We will call a linear functional satisfying condition (ii) of the preceding Proposition a *generating functional*. Lemma 1.5.7 and Proposition 1.5.8 show that Lévy processes can also be characterized by their generating functional $L = \frac{\mathrm{d}}{\mathrm{d}t}\Big|_{t=0} \varphi_t$.

1.5.3 The Schürmann triple of a Lévy process

Let D be a pre-Hilbert space. Then we denote by $\mathcal{L}(D)$ the set of all linear operators on D that have an adjoint defined everywhere on D, that is,

$$\mathcal{L}(D) = \left\{ X : D \to D \text{ linear} \;\middle|\; \begin{array}{l} \text{there exists } X^* : D \to D \text{ linear s.t.} \\ \langle u, Xv \rangle = \langle X^*u, v \rangle \text{ for all } u, v \in D \end{array} \right\}.$$

$\mathcal{L}(D)$ is clearly a unital $*$-algebra.

Definition 1.5.9 Let $\mathcal{B}$ be a unital $*$-algebra equipped with a unital hermitian character $\varepsilon : \mathcal{B} \to \mathbb{C}$ (that is, $\varepsilon(\mathbf{1}) = 1$, $\varepsilon(b^*) = \overline{\varepsilon(b)}$, and $\varepsilon(ab) = \varepsilon(a)\varepsilon(b)$ for all $a, b \in \mathcal{B}$). A *Schürmann triple on* $(\mathcal{B}, \varepsilon)$ is a triple (ρ, η, L) consisting of

- a unital $*$-representation $\rho : \mathcal{B} \to \mathcal{L}(D)$ of $\mathcal{B}$ on some pre-Hilbert space D,
- a ρ-ε-1-*cocycle* $\eta : \mathcal{B} \to D$, that is, a linear map $\eta : \mathcal{B} \to D$ such that

$$\eta(ab) = \rho(a)\eta(b) + \eta(a)\varepsilon(b) \tag{1.5.2}$$

  for all $a, b \in \mathcal{B}$, and
- a hermitian linear functional $L : \mathcal{B} \to \mathbb{C}$ that has the bilinear map $\mathcal{B} \times \mathcal{B} \ni (a, b) \mapsto -\langle \eta(a^*), \eta(b) \rangle$ as a ε-ε-2-*coboundary*, that is, that satisfies

$$-\langle \eta(a^*), \eta(b) \rangle = \partial L(a, b) = \varepsilon(a)L(b) - L(ab) + L(a)\varepsilon(b) \tag{1.5.3}$$

  for all $a, b \in \mathcal{B}$.

We will call a Schürmann triple *surjective*, if the cocycle $\eta : \mathcal{B} \to D$ is surjective.

Two surjective Schürmann triples (ρ_i, η_i, L_i), $i = 1, 2$, on a unital $*$-algebra $\mathcal{B}$ equipped with a character ε, defined over pre-Hilbert spaces D_1 and D_2, are said to be *unitarily equivalent*, if $L_1 = L_2$ and there exists a bijection $U : D_1 \to D_2$ such that $\rho_2(a)U = U\rho_1(a)$ and $\eta_2(a) = U\eta_1(a)$ for all $a \in \mathcal{B}$. The condition $L_1 = L_2$ guarantees that U is an isometry.

The functionals of two unitarily equivalent Schürmann triples have to coincide by definition, and, conversely, if the functionals of two surjective Schürmann triples agree, then the triples have to be unitarily equivalent. This can be shown in the same way as the uniqueness of the GNS construction.

Note that the bijection $U : D_1 \to D_2$ that implements the unitary equivalence extends to a unitary operator between the completions of D_1 and D_2.

Theorem 1.5.10 *Let $\mathcal{B}$ be an involutive bialgebra. We have one-to-one correspondences between Lévy processes on $\mathcal{B}$ (modulo equivalence), convolution semigroups of states on $\mathcal{B}$, generating functionals on $\mathcal{B}$, and surjective Schürmann triples on $\mathcal{B}$ (modulo unitary equivalence).*

Proof It only remains to establish the one-to-one correspondence between generating functionals and Schürmann triples.

Let (ρ, η, L) be a Schürmann triple, then we can show that L is a generating functional, that is, a hermitian, conditionally positive linear functional with $L(\mathbf{1}) = 0$.

The cocycle has to vanish on the unit element $\mathbf{1}$, since

$$\eta(\mathbf{1}) = \eta(\mathbf{1} \cdot \mathbf{1}) = \rho(\mathbf{1})\eta(\mathbf{1}) + \eta(\mathbf{1})\varepsilon(\mathbf{1}) = 2\eta(\mathbf{1}).$$

This implies

$$L(\mathbf{1}) = L(\mathbf{1} \cdot \mathbf{1}) = \varepsilon(\mathbf{1})L(\mathbf{1}) + \langle \eta(\mathbf{1}), \eta(\mathbf{1}) \rangle + L(\mathbf{1})\varepsilon(\mathbf{1}) = 2L(\mathbf{1}) = 0.$$

Furthermore, L is hermitian by definition and conditionally positive, since by (1.5.3) we get

$$L(b^*b) = \langle \eta(b), \eta(b) \rangle = \|\eta(b)\|^2 \geq 0$$

for $b \in \ker \varepsilon$.

Let now L be a generating functional. The sesqui-linear form $\langle\cdot,\cdot\rangle_L : \mathcal{B}\times\mathcal{B}\to\mathbb{C}$ defined by

$$\langle a,b\rangle_L = L\Big(\big(a-\varepsilon(a)\mathbf{1}\big)^*\big(b-\varepsilon(b)\mathbf{1}\big)\Big)$$

for $a,b\in\mathcal{B}$ is positive, since L is conditionally positive. Dividing $\mathcal{B}$ by the null space

$$\mathcal{N}_L = \{a\in\mathcal{B}|\langle a,a\rangle_L = 0\}$$

we obtain a pre-Hilbert space $D=\mathcal{B}/\mathcal{N}_L$ with a positive definite inner product $\langle\cdot,\cdot\rangle$ induced by $\langle\cdot,\cdot\rangle_L$. For the cocycle $\eta:\mathcal{B}\to D$ we take the canonical projection, this is clearly surjective and satisfies Equation (1.5.3).

The $*$-representation ρ is induced from the left multiplication on $\mathcal{B}$ on $\ker\varepsilon$, that is,

$$\rho(a)\eta\big(b-\varepsilon(b)\mathbf{1}\big)=\eta\Big(a\big(b-\varepsilon(b)\mathbf{1}\big)\Big) \text{ or } \rho(a)\eta(b)=\eta(ab)-\eta(a)\varepsilon(b)$$

for $a,b\in\mathcal{B}$. To show that this is well defined, we have to verify that left multiplication by elements of $\mathcal{B}$ leaves the null-space invariant. Let therefore $a,b\in\mathcal{B}$, $b\in\mathcal{N}_L$, then we have

$$\begin{aligned}\Big|\Big|\Big(a\big(b-\varepsilon(b)\mathbf{1}\big)\Big)\Big|\Big|^2 &= L\Big(\big(ab-a\varepsilon(b)\mathbf{1}\big)^*\big(ab-a\varepsilon(b)\mathbf{1}\big)\Big)\\ &= L\Big(\big(b-\varepsilon(b)\mathbf{1}\big)^*a^*\big(ab-a\varepsilon(b)\mathbf{1}\big)\Big)\\ &= \big\langle b-\varepsilon(b)\mathbf{1}, a^*a\big(b-\varepsilon(b)\mathbf{1}\big)\big\rangle_L\\ &\le ||b-\varepsilon(b)\mathbf{1}||^2\ \big|\big|a^*a\big(b-\varepsilon(b)\mathbf{1}\big)\big|\big|^2 = 0,\end{aligned}$$

with Schwarz' inequality.

That the Schürmann triple (ρ,η,L) obtained in this way is unique up to unitary equivalence follows similarly as for the usual GNS construction. □

Exercise 1.5.11 Let $(X_t)_{t\ge 0}$ be a classical real-valued Lévy process with all moments finite (on some probability space $(\Omega,\mathcal{F},P)$). Define a Lévy process on the free unital algebra $\mathbb{C}[x]$ generated by one symmetric element $x=x^*$ with the coproduct and counit determined by $\Delta(x)=x\otimes\mathbf{1}+\mathbf{1}\otimes x$ and $\varepsilon(x)=0$, such that

$$\Phi\big(j_{st}(x^k)\big)=\mathbb{E}\Big(\big(X_t-X_s\big)^k\Big)$$

holds for all $k \in \mathbb{N}$ and all $0 \le s \le t$.

Construct the Schürmann triple for Brownian motion and for a compound Poisson process (with finite moments).

For the classification of Gaussian and drift generating functionals on an involutive bialgebra $\mathcal{B}$ with counit ε, we need the ideals

$$\begin{aligned} K &= \ker \varepsilon, \\ K^2 &= \operatorname{span}\{ab | a, b \in K\}, \\ K^3 &= \operatorname{span}\{abc | a, b, c \in K\}. \end{aligned}$$

Proposition 1.5.12 *Let L be a generating functional on $\mathcal{B}$ with Schürmann triple (ρ, η, L). Then the following are equivalent.*

(i) $\eta = 0$,
(ii) $L|_{K^2} = 0$,
(iii) *L is an ε-derivation, that is,* $L(ab) = \varepsilon(a)L(b) + L(a)\varepsilon(b)$ *for all* $a, b \in \mathcal{B}$,
(iv) *The states φ_t are multiplicative, that is,* $\varphi_t(ab) = \varphi_t(a)\varphi_t(b)$ *for all* $a, b \in \mathcal{B}$ *and* $t \ge 0$.

If a generating functional L satisfies one of these conditions, then we call it and the associated Lévy process a *drift*.

Proposition 1.5.13 *Let L be a generating functional on $\mathcal{B}$.*
Then the following are equivalent.

(i) $L|_{K^3} = 0$,
(ii) $L(b^*b) = 0$ *for all* $b \in K^2$,
(iii) $L(abc) = L(ab)\varepsilon(c) + L(ac)\varepsilon(b) + L(bc)\varepsilon(a) - \varepsilon(ab)L(c) - \varepsilon(ac)L(b) - \varepsilon(bc)L(a)$ *for all* $a, b, c \in \mathcal{B}$,
(iv) $\rho|_K = 0$ *for the representation ρ in the surjective Schürmann triple (ρ, η, L) associated to L by the GNS-type construction presented in the proof of Theorem 1.5.10,*
(v) $\rho = \varepsilon\mathbf{1}$, *for the representation ρ in the surjective Schürmann triple (ρ, η, L) associated to L by the GNS-type construction presented in the proof of Theorem 1.5.10,*
(vi) $\eta|_{K^2} = 0$ *for the cocycle η in any Schürmann triple (ρ, η, L) containing L,*
(vii) $\eta(ab) = \varepsilon(a)\eta(b) + \eta(a)\varepsilon(b)$ *for all* $a, b \in \mathcal{B}$ *and the cocycle η in any Schürmann triple (ρ, η, L) containing L.*

If a generating functional L satisfies one of these conditions, then we call it and also the associated Lévy process *quadratic* or *Gaussian*.

The proofs of the preceding two propositions can be carried out as an exercise or found in [Sch93, Section 5.1].

Proposition 1.5.14 *Let L be a generating functional on $\mathcal{B}$. Then the following are equivalent.*

(i) There exists a state $\varphi : \mathcal{B} \to \mathbb{C}$ and a real number $\lambda > 0$ such that

$$L(b) = \lambda\big(\varphi(b) - \varepsilon(b)\big)$$

for all $b \in \mathcal{B}$.

(ii) There exists a Schürmann triple (ρ, η, L) containing L, in which the cocycle η is trivial, that is, of the form

$$\eta(b) = \big(\rho(b) - \varepsilon(b)\big)\omega, \qquad \text{for all } b \in \mathcal{B},$$

for some non-zero vector $\omega \in D$. In this case we will also call η the coboundary *of the vector ω.*

If a generating functional L satisfies one of these conditions, then we call it a *Poisson generating functional* and the associated Lévy process a *compound Poisson process*.

Proof To show that (ii) implies (i), set $\varphi(b) = \frac{\langle \omega, \rho(b)\omega\rangle}{\langle \omega, \omega\rangle}$ and $\lambda = \|\omega\|^2$.

For the converse, let (D, ρ, ω) be the GNS triple for $(\mathcal{B}, \varphi)$ and check that (ρ, η, L) with $\eta(b) = \big(\rho(b) - \varepsilon(b)\big)\omega$, $b \in \mathcal{B}$ defines a Schürmann triple. □

Remark 1.5.15 The Schürmann triple for a Poisson generating functional $L = \lambda(\varphi - \varepsilon)$ obtained by the GNS construction for φ is not necessarily surjective. Consider, e.g., a classical additive $\mathbb{R}$-valued compound Poisson process, whose Lévy measure μ is not supported on a finite set. Then the construction of a surjective Schürmann triple in the proof of Theorem 1.5.10 gives the pre-Hilbert space $D_0 = \operatorname{span}\{x^k | k = 1, 2, \ldots\} \subseteq L^2(\mathbb{R}, \mu)$. On the other hand, the GNS construction for φ leads to the pre-Hilbert space $D = \operatorname{span}\{x^k | k = 0, 1, 2, \ldots\} \subseteq L^2(\mathbb{R}, \mu)$. The cocycle η is the coboundary of the constant function, which is not contained in D_0.

1.5.4 Examples

Lévy processes on the circle $\mathbb{T}$ and on the real line $\mathbb{R}$
Consider the involutive bialgebra

$$\mathcal{B} = \text{span}\{e_\lambda; \lambda \in \mathbb{R}\}$$

with the multiplication

$$e_\lambda \cdot e_\mu = e_{\lambda+\mu}, \qquad \lambda, \mu \in \mathbb{R},$$

involution $e_\lambda^* = e_{-\lambda}$ for $\lambda \in \mathbb{R}$, coproduct

$$\Delta(e_\lambda) = e_\lambda \otimes e_\lambda$$

and counit $\varepsilon(e_\lambda) = 1$ for all $\lambda, \mu \in \mathbb{R}$. Consider also the subalgebras

$$\mathcal{B}_\alpha = \text{span}\{e_{k\alpha}; k \in \mathbb{Z}\}$$

for $\alpha > 0$.

The basis elements e_λ can be represented as exponential functions $e_\lambda : \mathbb{R} \ni x \to e^{i\lambda x} \in \mathbb{C}$. As this representation of $\mathcal{B}$ is faithful, we can view $\mathcal{B}$ as a subalgebra of the algebra $\mathcal{R}(\mathbb{R})$ of representative functions on $\mathbb{R}$.

The subalgebra $\mathcal{B}_\alpha$ is generated as a $*$-algebra by one unitary element e_α; it is therefore commutative and isomorphic to the algebra of polynomials on the unit circle $\mathbb{T}$.

Lévy processes on $\mathcal{R}(\mathbb{R})$, $\mathcal{B}$, and $\mathcal{B}_\alpha$ can be constructed from real-valued Lévy processes. If $(X_t)_{t\geq 0}$ is a real-valued Lévy process, then we would like to define a Lévy process on $f \in \mathcal{R}(\mathbb{R})$ by

$$j_{st}(f) = f(X_t - X_s)$$

for $0 \leq s \leq t$ and $f \in \mathcal{R}(\mathbb{R})$, but since the expectation of $f(X_t - X_s)$ might not be defined, we have to be more careful. We can do this by restricting to appropriate subalgebras of $\mathcal{R}(\mathbb{R})$. E.g., if we restrict the homomorphisms $(j_{st})_{0\leq s\leq t}$ to $\mathcal{B}$ or $\mathcal{B}_\alpha$, then

$$j_{st}(e_\lambda) = \exp\big(i\lambda(X_t - X_s)\big)$$

are bounded random variables defined on the same probability space $(\Omega, \mathcal{F}, P)$ as the Lévy process $(X_t)_{t\geq 0}$, so we can view $(j_{st})_{0\leq s\leq t}$ as a Lévy process on the involutive bialgebras $\mathcal{B}$ or $\mathcal{B}_\alpha$

over the algebraic probability space $(L^\infty(\Omega,\mathcal{F},P),\mathbb{E})$, where $\mathbb{E}$ denotes expectation w.r.t. P.

If the Lévy process $(X_t)_{t\geq 0}$ has finite moments, then we can restrict similarly to polynomials on $\mathbb{R}$ and we get a Lévy process *over* the algebraic probability space $(L^{\infty-}(\Omega,\mathcal{F},P),\mathbb{E})$, with

$$L^{\infty-}(\Omega,\mathcal{F},P)=\bigcap_{1\leq p<\infty} L^p(\Omega,\mathcal{F},P).$$

Lévy processes on finite semigroups

Exercise 1.5.16 Let $(G,\cdot,e)$ be a finite semigroup with unit element e. Then the complex-valued functions $\mathcal{F}(G)$ on G form an involutive bialgebra. The algebra structure and the involution are given by pointwise multiplication and complex conjugation. The coproduct and counit are defined by

$$\begin{aligned}\Delta(f)(g_1,g_2) &= f(g_1\cdot g_2) \qquad \text{for } g_1,g_2\in G,\\ \varepsilon(f) &= f(e),\end{aligned}$$

for $f\in\mathcal{F}(G)$.

Show that the classical Lévy processes in G are in one-to-one correspondence with Lévy processes on the $*$-bialgebra $\mathcal{F}(G)$.

1.6 Lévy Processes on Compact Quantum Groups and their Markov Semigroups

In the eighties of the last century Woronowicz introduced 'compact matrix pseudogroups', which are nowadays called compact quantum groups, *cf.* [Wor87a, Wor87b, Wor88]. They can be defined alternatively as C*-bialgebras, that is, unital C*-algebras equipped with additional structure, see Definition 1.6.1 below, which satisfy the quantum cancellation properties, or as involutive Hopf algebras satisfying additional conditions, see [DK94]. For more background on compact quantum groups, see, for example, [Wor98, MvD98, MT04, Tim08, NT13].

The additional structure makes compact quantum groups correspond more closely to compact groups than arbirtrary involutive bialgebras, see Subsection 1.6.1 below. This leads to a richer theory for Lévy processes on these structures, which we will present in this Section.

We will show that Lévy processes on compact quantum groups can be characterized as time- and space-homogeneous Markov process, just like in classical probability, see Theorem 1.6.6. This observation is the starting point of [CFK14], where other symmetry properties, like GNS- or KMS-symmetry or invariance under the adjoint action, of the Markov semigroups associated to Lévy processes on compact quantum groups are studied. It is possible to associate Dirichlet forms, derivations, and Dirac operators to these Markov semigroups. We refer the interested reader to the paper [CFK14] for more information. See also the article [CS15] by Caspers and Skalski for an application of Dirichlet forms on compact quantum groups to the Haagerup approximation property.

In this Section, we will work both with algebraic and topologic dual spaces and tensor products. We use the following conventions. For an algebra $\mathcal{A}$, if no norm or topology is specified, $\mathcal{A}'$ will denote the algebraic dual of $\mathcal{A}$, that is, the space of all linear functionals from $\mathcal{A}$ to $\mathbb{C}$. For a C^*-algebra A, by A' we will mean the dual space consisting of all linear continuous functionals from A to $\mathbb{C}$. The symbol $\otimes$ will denote the spatial tensor product when it is used for C^*-algebras and the algebraic tensor product otherwise. We will never need algebraic tensor products of C^*-algebras. See, e.g., [Ped89] and the lecture 'Quantum dynamical systems from the point of view of non commutative mathematics' by Adam Skalski for the spatial tensor product and other facts about C^*-algebras.

1.6.1 Compact quantum groups

The notion of compact quantum groups has been introduced in [Wor87a]. Here we adopt the definition from [Wor98] (Definition 1.1 of that paper).

Definition 1.6.1 A *C^*-bialgebra* (a compact quantum semigroup) is a pair (A, Δ), where A is a unital C^*-algebra, $\Delta : \mathsf{A} \to \mathsf{A} \otimes \mathsf{A}$ is a unital, $*$-homomorphic map which is coassociative, that is,

$$(\Delta \otimes \mathrm{id}_{\mathsf{A}}) \circ \Delta = (\mathrm{id}_{\mathsf{A}} \otimes \Delta) \circ \Delta.$$

If the *quantum cancellation properties*

$$\overline{\mathrm{Lin}}((1 \otimes \mathsf{A})\Delta(\mathsf{A})) = \overline{\mathrm{Lin}}((\mathsf{A} \otimes 1)\Delta(\mathsf{A})) = \mathsf{A} \otimes \mathsf{A},$$

are satisfied, then the pair (A, Δ) is called a *compact quantum group* (CQG).

If the algebra A of a compact quantum group is commutative, then A is isomorphic to the algebra $C(G)$ of continuous functions on a compact group G. To emphasize that for an arbitrary (that is, not necessarily noncommutative) compact quantum group (A, Δ) the algebra A replaces the algebra of continuous functions on an (abstract) quantum analog of a group, the notation $\mathbb{G} = (\mathsf{A}, \Delta)$ and $\mathsf{A} = C(\mathbb{G})$ is also frequently used.

The map Δ is called the *coproduct* of A and it induces the *convolution product* of functionals

$$\lambda \star \mu := (\lambda \otimes \mu) \circ \Delta, \quad \lambda, \mu \in \mathsf{A}'.$$

The following fact is of the fundamental importance, *cf.* [Wor98, Theorem 2.3].

Proposition 1.6.2 *Let* A *be a compact quantum group. There exists a unique state* $h \in \mathsf{A}'$ *(called the* Haar state *of* A*) such that for all* $a \in \mathsf{A}$

$$(h \otimes \mathrm{id}_{\mathsf{A}}) \circ \Delta(a) = (\mathrm{id}_{\mathsf{A}} \otimes h) \circ \Delta(a) = h(a)1.$$

In general, the Haar state of a compact quantum group need not be faithful or tracial.

Corepresentations

An element $u = (u_{jk})_{1 \leq j,k \leq n} \in M_n(\mathsf{A})$ is called an *n-dimensional corepresentation of* $\mathbb{G} = (\mathsf{A}, \Delta)$ if for all $j, k = 1, \ldots, n$ we have $\Delta(u_{jk}) = \sum_{p=1}^{n} u_{jp} \otimes u_{pk}$. All corepresentations considered in this course are supposed to be finite-dimensional. A corepresentation u is said to be *non-degenerate,* if u is invertible, *unitary,* if u is unitary, and *irreducible,* if the only matrices $T \in M_n(\mathbb{C})$ with $Tu = uT$ are multiples of the identity matrix. Two corepresentations $u, v \in M_n(\mathsf{A})$ are called *equivalent,* if there exists an invertible matrix $U \in M_n(\mathbb{C})$ such that $Uu = vU$.

An important feature of compact quantum groups is the existence of a dense $*$-subalgebra $\mathcal{A}$ (the algebra of the *polynomials* of A), which is in fact a Hopf $*$-algebra – so, for example $\Delta : \mathcal{A} \to \mathcal{A} \otimes \mathcal{A}$. With the notation $\mathbb{G} = (\mathsf{A}, \Delta)$, one often refers to $\mathcal{A}$ as $\mathrm{Pol}(\mathbb{G})$.

Fix a complete family $(u^{(s)})_{s \in \mathcal{I}}$ of mutually inequivalent irreducible unitary corepresentations of A, then $\{u^{(s)}_{k\ell}; s \in \mathcal{I}, 1 \leq k,$

$\ell \leq n_s\}$ (where n_s denotes the dimension of $u^{(s)}$) is a linear basis of $\mathrm{Pol}(\mathbb{G})$, *cf.* [Wor98, Proposition 5.1]. We shall reserve the index $s = 0$ for the trivial corepresentation $u^{(0)} = \mathbf{1}$. The Hopf algebra structure on $\mathrm{Pol}(\mathbb{G})$ is defined by

$$\varepsilon(u_{jk}^{(s)}) = \delta_{jk}, \quad S(u_{jk}^{(s)}) = (u_{kj}^{(s)})^* \quad \text{for } j,k = 1,\ldots,n_s,$$

where $\varepsilon : \mathrm{Pol}(\mathbb{G}) \to \mathbb{C}$ is the counit and $S : \mathrm{Pol}(\mathbb{G}) \to \mathrm{Pol}(\mathbb{G})$ is the antipode. They satisfy

$$(\mathrm{id} \otimes \varepsilon) \circ \Delta = \mathrm{id} = (\varepsilon \otimes \mathrm{id}) \circ \Delta, \tag{1.6.1}$$

$$m_{\mathcal{A}} \circ (\mathrm{id} \otimes S) \circ \Delta = \varepsilon(a)\mathbf{1} = m_{\mathcal{A}} \circ (S \otimes \mathrm{id}) \circ \Delta, \tag{1.6.2}$$

$$\big(S(a^*)^*\big) = a \tag{1.6.3}$$

for all $a \in \mathrm{Pol}(\mathbb{G})$. Let us also remind that the Haar state is always faithful on $\mathrm{Pol}(\mathbb{G})$.

Set $V_s = \mathrm{span}\{u_{jk}^{(s)}; 1 \leq j,k \leq n_s\}$ for $s \in \mathcal{I}$. Let $u^{(s)}$ be an irreducible unitary corepresentation of $\mathbb{G}$. Then $\overline{u^{(s)}} = \big(u_{jk}^{(s)*}\big)_{1\leq j,k\leq n_s}$ is also an irreducible non-degenerate corepresentation, but no longer unitary. By [Wor98, Proposition 5.2], there exists an irreducible unitary corepresentation $u^{(s^c)}$, called the *contragredient* corepresentation of $u^{(s)}$, that is equivalent to $\overline{u^{(s)}}$. We have $V_s^* = V_{s^c}$ and $(s^c)^c = s$.

The dual discrete quantum group

To every compact quantum group $\mathbb{G} = (\mathsf{A}, \Delta)$ there exists a dual discrete quantum group $\hat{\mathbb{G}}$, *cf.* [PW90]. For our purposes it will be most convenient to introduce $\hat{\mathbb{G}}$ in the setting of Van Daele's algebraic quantum groups, *cf.* [VD98, VD03]. However, the reader should be aware that we adopt a slightly different convention for the Fourier transform.

A pair (A, Δ), consisting of a $*$-algebra A (with or without identity) and a coassociative comultiplication $\Delta : A \to M(A \otimes A)$, is called an *algebraic quantum group* if the product is non-degenerate (that is, $ab = 0$ for all a implies $b = 0$), if the two operators $T_1 : A \otimes A \ni a \otimes b \mapsto \Delta(a)(b \otimes 1) \in A \otimes A$ and $T_2 : A \otimes A \ni a \otimes b \mapsto \Delta(a)(1 \otimes b) \in A \otimes A$ are well defined bijections and if there exists a non-zero left invariant positive functional on A. Here, $M(B)$ denotes the set of multipliers on B of a $*$-algebra B, and by well definedness we mean that $\Delta(a)(b \otimes 1)$ and $\Delta(a)(1 \otimes b)$ should belong to $A \otimes A$ for all $a, b \in A$ (viewed as a subalgebra of $M(A \otimes A)$). We refer the reader to [VD98] for details.

If (A, Δ) is a compact quantum group, then $(\mathrm{Pol}(\mathbb{G}), \Delta|_{\mathcal{A}})$ is an algebraic quantum group (of compact type) and the Haar state is a faithful left and right integral.

For $a \in \mathrm{Pol}(\mathbb{G})$ we can define $h_a \in \mathrm{Pol}(\mathbb{G})'$ by the formula

$$h_a(b) = h(ab) \quad \text{for } b \in \mathrm{Pol}(\mathbb{G}),$$

where h is the Haar state, and we denote by $\hat{\mathcal{A}}$ the space of linear functionals on $\mathcal{A} = \mathrm{Pol}(\mathbb{G})$ of the form h_a for $a \in \mathrm{Pol}(\mathbb{G})$.

The set $\hat{\mathcal{A}}$ becomes an associative $*$-algebra with the convolution of functionals as the multiplication: $\lambda \star \mu = (\lambda \otimes \mu) \circ \Delta$, and the involution $\lambda^*(x) = \overline{\lambda(S(x)^*)}$ $(\lambda, \mu \in \hat{\mathcal{A}})$. The Hopf structure is given as follows: the coproduct $\hat{\Delta}$ is the dual of the product on $\mathrm{Pol}(\mathbb{G})$, the antipode $\hat{S}$ is the dual to S and the counit $\hat{\varepsilon}$ is the evaluation in $\mathbf{1}$. In particular, we have $\hat{S}(\lambda)(x) = \lambda(Sx)$ for $\lambda \in \hat{\mathcal{A}}$, $x \in \mathrm{Pol}(\mathbb{G})$ and if $\hat{\Delta}(\lambda) \in \hat{\mathcal{A}} \otimes \hat{\mathcal{A}}$ then

$$\hat{\Delta}(\lambda)(x \otimes y) = \lambda_{(1)}(x) \otimes \lambda_{(2)}(y) = \lambda(xy), \quad x, y \in \mathrm{Pol}(\mathbb{G}).$$

The pair $\hat{\mathbb{G}} = (\hat{\mathcal{A}}, \hat{\Delta})$ is an algebraic quantum group, called the *dual* of $\mathbb{G}$.

The linear map that associates to $a \in \mathrm{Pol}(\mathbb{G})$ the functional $h_a \in \hat{\mathcal{A}}$ is called the *Fourier transform* and its value on an element a is also denoted by $\hat{a}$. Let us note that, due to the faithfulness of the Haar state h, $\hat{\mathcal{A}}$ separates the points of $\mathrm{Pol}(\mathbb{G})$.

Woronowicz characters and modular automorphism group

For $a \in \mathsf{A}$, $\lambda \in \mathsf{A}'$ we define

$$\begin{aligned} \lambda \star a &= (\mathrm{id} \otimes \lambda)\Delta(a), \\ a \star \lambda &= (\lambda \otimes \mathrm{id})\Delta(a). \end{aligned}$$

If $a \in \mathrm{Pol}(\mathbb{G})$ and $\lambda \in \mathrm{Pol}(\mathbb{G})'$, then $\lambda \star a, a \star \lambda \in \mathrm{Pol}(\mathbb{G})$.

For a compact quantum group A with the dense $*$-Hopf algebra $\mathrm{Pol}(\mathbb{G})$, there exists a unique family $(f_z)_{z \in \mathbb{C}}$ of linear multiplicative functionals on $\mathrm{Pol}(\mathbb{G})$, called *Woronowicz characters* (*cf.* [Wor98, Theorem 1.4]), such that

(i) $f_z(\mathbf{1}) = 1$ for $z \in \mathbb{C}$,

(ii) the mapping $\mathbb{C} \ni z \mapsto f_z(a) \in \mathbb{C}$ is an entire holomorphic function for all $a \in \mathrm{Pol}(\mathbb{G})$,

(iii) $f_0(z) = \varepsilon$ and $f_{z_1} \star f_{z_2} = f_{z_1+z_2}$ for any $z_1, z_2 \in \mathbb{C}$,

(iv) $f_z(S(a)) = f_{-z}(a)$ and $f_{\bar{z}}(a^*) = \overline{f_{-z}(a)}$ for any $z \in \mathbb{C}$, $a \in \mathrm{Pol}(\mathbb{G})$,

(v) $S^2(a) = f_{-1} \star a \star f_1$ for $a \in \mathrm{Pol}(\mathbb{G})$,

(vi) the Haar state h satisfies:

$$h(ab) = h(b(f_1 \star a \star f_1)), \quad a, b \in \mathrm{Pol}(\mathbb{G}).$$

In this case the formula

$$\sigma_t(a) = f_{it} \star a \star f_{it}, \quad t \geq 0, \tag{1.6.4}$$

defines a one-parameter group of modular automorphisms of $\mathrm{Pol}(\mathbb{G})$ and h is the $(\sigma, -1)$-KMS state, which means that it satisfies

$$h(ab) = h(b\sigma_{-i}(a)), \quad a, b \in \mathrm{Pol}(\mathbb{G}), \tag{1.6.5}$$

cf. [BR97, Definition 5.3.1]. Such a state is σ-invariant, that is, $h(\sigma_t(a)) = h(a)$ for $a \in \mathrm{Pol}(\mathbb{G})$ and $t \geq 0$ (see [BR97, Proposition 5.3.3]).

The matrix elements of the irreducible unitary corepresentations satisfy the famous generalized Peter–Weyl orthogonality relations

$$h\left(\left(u_{ij}^{(s)}\right)^* u_{k\ell}^{(t)}\right) = \frac{\delta_{st}\delta_{j\ell} f_{-1}\left(u_{ki}^{(s)}\right)}{D_s}, \quad h\left(u_{ij}^{(s)}\left(u_{k\ell}^{(t)}\right)^*\right) = \frac{\delta_{st}\delta_{ik} f_1\left(u_{\ell j}^{(s)}\right)}{D_s}, \tag{1.6.6}$$

where $f_1 : \mathrm{Pol}(\mathbb{G}) \to \mathbb{C}$ is the Woronowicz character and

$$D_s = \sum_{\ell=1}^{n_s} f_1\left(u_{\ell\ell}^{(s)}\right)$$

is the quantum dimension of $u^{(s)}$, *cf.* [Wor87a, Theorem 5.7.4]. Note that unitarity implies that the matrix

$$\left(f_1\left(\left(u_{jk}^{(s)}\right)^*\right)\right)_{1 \leq j,k \leq n_s} \in M_{n_s}(\mathbb{C})$$

is invertible, with inverse $\left(f_1(u^{(s)}_{jk})\right)_{jk} \in M_{n_s}(\mathbb{C})$, *cf.* [Wor87a, Equation (5.24)].

Remark 1.6.3 The Haar state on compact quantum groups is a trace if and only if the antipode is involutive, that is, we have $S^2(a) = a$ for all $a \in \mathrm{Pol}(\mathbb{G})$. In this case we say that (A, Δ) is of *Kac type*. This is also equivalent to the following conditions, *cf.* [Wor98, Theorem 1.5],

(i) $f_z = \varepsilon$ for all $z \in \mathbb{C}$,

(ii) $\sigma_t = \mathrm{id}$ for all $t \in \mathbb{R}$.

The antipode $S : \mathrm{Pol}(\mathbb{G}) \to \mathrm{Pol}(\mathbb{G})$ is a closable operator and its closure $\overline{S}$ admits a polar decomposition

$$\overline{S} = R \circ \tau_{\frac{i}{2}}, \tag{1.6.7}$$

where $\tau_{\frac{i}{2}}$ is the analytic generator of a one-parameter group $(\tau_t)_{t\in\mathbb{R}}$ of $*$-automorphisms of the C*-algebra A and $R : \mathsf{A} \to \mathsf{A}$ is a linear antimultiplicative norm preserving involution that commutes with hermitian conjugation and with the semigroup (τ_t), that is, $\tau_t \circ R = R \circ \tau_t$ for all $t \in \mathbb{R}$, see [Wor98, Theorem 1.6]. The operator R is called the *unitary antipode* and the one-parameter group $(\tau_t)_{t\in\mathbb{R}}$ is called the *scaling group*.

Moreover, τ and R are related to Woronowicz characters through the following formulae

$$\tau_t(a) = f_{it} \star a \star f_{-it}, \tag{1.6.8}$$

$$R(a) = S(f_{\frac{1}{2}} \star a \star f_{-\frac{1}{2}}) \tag{1.6.9}$$

for $a \in \mathrm{Pol}(\mathbb{G})$.

1.6.2 Translation invariant Markov semigroups

Our goal is to construct Markov semigroups on compact quantum groups that reflect the structure of the quantum group. In this section we show that it is exactly the *translation invariant* Markovian semigroups that can be obtained from Lévy processes on the algebra of smooth functions $\mathcal{A} = \mathrm{Pol}(\mathbb{G})$ of the quantum group $\mathbb{G} = (\mathsf{A}, \Delta)$.

For this purpose, we first prove that the Markov semigroup $(T_t)_{t\geq 0}$ of a Lévy process on $\mathrm{Pol}(\mathbb{G})$ has a unique extension to a strongly continuous Markov semigroup on both its reduced and its universal C*-algebra. We then show that the characterization of Lévy processes in topological groups as the Markov processes that are invariant under time and space translations extends to compact quantum groups.

If $(j_{st})_{0\leq s\leq t}$ is a Lévy process on an $*$-algebra $\mathrm{Pol}(\mathbb{G})$ with the convolution semigroup of states $(\varphi_t)_{t\geq 0}$ on $\mathrm{Pol}(\mathbb{G})$ and the Markov semigroup $(T_t)_{t\geq 0}$ on $\mathrm{Pol}(\mathbb{G})$, then, by universality, each φ_t extends to a continuous functional on the universal C*-algebra $C_u(\mathbb{G})$ generated by $\mathrm{Pol}(\mathbb{G})$ (see also [BMT01, Theorem 3.3] for a detailed exposition). In the literature this C*-algebra is also denoted by A_u. Then the formula $T_t = (\mathrm{id}\otimes\varphi_t)\circ\Delta$ makes sense on $C_u(\mathbb{G})$ (where $\Delta : C_u(\mathbb{G}) \to C_u(\mathbb{G})\otimes C_u(\mathbb{G})$ denotes the unique unital $*$-homomorphism that extends $\Delta : \mathrm{Pol}(\mathbb{G}) \to \mathrm{Pol}(\mathbb{G})\otimes \mathrm{Pol}(\mathbb{G})$) and one easily shows (in the same way as in the Proposition below) that $(T_t)_t$ becomes a strongly continuous Markov semigroup of contractions on $C_u(\mathbb{G})$. This means that each T_t ($t\geq 0$) is a unital, completely positive contraction and $(T_t)_t$ is a strongly continuous semigroup on $C_u(\mathbb{G})$.

For us, however, it will be more natural to consider the reduced C*-algebra generated by $\mathrm{Pol}(\mathbb{G})$. This is the C*-algebra $\mathsf{A}_r = C_r(\mathbb{G})$ obtained by taking the norm closure of the GNS representation of $\mathrm{Pol}(\mathbb{G})$ with respect to the Haar state h. The Haar state h is by construction faithful on $C_r(\mathbb{G})$. The coproduct on $\mathrm{Pol}(\mathbb{G})$ extends to a unique unital $*$-homomorphism $\Delta : C_r(\mathbb{G}) \to C_r(\mathbb{G})\otimes C_r(\mathbb{G})$, which makes the pair $(C_r(\mathbb{G}),\Delta)$ a compact quantum group. The following result shows that, even though $\varphi_t : \mathrm{Pol}(\mathbb{G}) \to \mathbb{C}$ can be unbounded with respect to the reduced C*-norm and therefore not extend to $C_r(\mathbb{G})$, $(T_t)_t$ always extends to a strongly continuous Markov semigroup on $C_r(\mathbb{G})$.

States on any C*-algebraic version $C(\mathbb{G})$ of $\mathbb{G}$ define continuous convolution operators on the reduced version $C_r(\mathbb{G})$, *cf.* [Bra12, Lemma 3.4]. We will need a similar result for convolution semigroups of states on $\mathrm{Pol}(\mathbb{G})$.

Theorem 1.6.4 *Each Lévy process $(j_{st})_{0\leq s\leq t}$ on the Hopf $*$-algebra* $\mathrm{Pol}(\mathbb{G})$ *gives rise to a unique strongly continuous Markov semigroup $(T_t)_{t\geq 0}$ on $C_r(\mathbb{G})$, the reduced C*-algebra generated by* $\mathrm{Pol}(\mathbb{G})$.

This follows from the fact that the coproduct $\Delta : \mathrm{Pol}(\mathbb{G}) \to \mathrm{Pol}(\mathbb{G})\otimes \mathrm{Pol}(\mathbb{G})$ extends to a $*$-homomorphism $C_r(\mathbb{G}) \to C_r(\mathbb{G})\otimes$

$C_u(\mathbb{G})$, *cf.* [Wor88, BMT01]. This fact can be deduced from the properties of the right regular representation, see [Wor98], or also [MvD98, Section 5], [Kus97, Theorem 7.9].

We give here a detailed self-contained proof that uses explicitly the invariance of the Haar state.

Proof Let $\mathcal{H}$ be the Hilbert space of the GNS representation of $\mathrm{Pol}(\mathbb{G})$ associated to the Haar state h and let ξ denotes the related (normalized) cyclic vector. Then for any $a \in \mathrm{Pol}(\mathbb{G})$ we have $h(a) = \langle \xi, \lambda(a)\xi \rangle$, where λ is the left regular representation. We denote by $\|\cdot\|_r$ the norm in $C_r(\mathbb{G})$, that is $\|a\|_r = \|\lambda(a)\|$, where $\|\cdot\|$ denotes the operator norm.

In a similar way, we associate the Hilbert space $\mathcal{H}_t$, the GNS representation ρ_t on $\mathcal{H}_t$ and the normalized cyclic vector ξ_t to each state φ_t ($\varphi_t = \Phi \circ j_{0t}$, *cf.* Definition 1.5.3). We have $\varphi_t(a) = \langle \xi_t, \rho_t(a)\xi_t \rangle$ for $a \in \mathrm{Pol}(\mathbb{G})$.

We define the operators

$$\begin{aligned} i_t &: \mathcal{H} \ni v \to v \otimes \xi_t \in \mathcal{H} \otimes \mathcal{H}_t \\ \pi_t &: \mathcal{H} \otimes \mathcal{H}_t \ni v \otimes w \to \langle \xi_t, w \rangle_{\mathcal{H}_t} v \in \mathcal{H} \\ E_t &: B(\mathcal{H} \otimes \mathcal{H}_t) \ni X \to \pi_t \circ X \circ i_t \in B(\mathcal{H}). \end{aligned}$$

Since for each t, i_t is an isometry and π_t is contractive, E_t is contractive too: $\|E_t(X)\| = \|\pi_t \circ X \circ i_t\| \le \|X\|$.

Next we define

$$\begin{aligned} U &: \lambda(\mathrm{Pol}(\mathbb{G}))\xi \otimes \rho_t(\mathrm{Pol}(\mathbb{G}))\xi_t \to \mathcal{H} \otimes \mathcal{H}_t \\ &\lambda(a)\xi \otimes \rho_t(b)\xi_t \mapsto \lambda(a_{(1)})\xi \otimes \rho_t(a_{(2)}b)\xi_t \end{aligned}$$

and we check that it is an isometry with adjoint given by

$$U^*(\lambda(a)\xi \otimes \rho_t(b)\xi_t) = \lambda(a_{(1)})\xi \otimes \rho_t(S(a_{(2)})b)\xi_t.$$

Indeed, using the invariance of the Haar measure, we show that U is isometric

$$\begin{aligned} &\|U(\lambda(a)\xi \otimes \rho_t(b)\xi_t)\|^2 = \|\lambda(a_{(1)})\xi \otimes \rho_t(a_{(2)}b)\xi_t\|^2 \\ &\quad = h(a^*_{(1)}a_{(1)})\varphi_t(b^*a^*_{(2)}a_{(2)}b) = (h \otimes \varphi_t^b)(a^*_{(1)}a_{(1)} \otimes a^*_{(2)}a_{(2)}) \\ &\quad = (h \star \varphi_t^b)(a^*a) = h(a^*a)\varphi_t^b(\mathbf{1}) = h(a^*a)\varphi_t(b^*b) = \|\lambda(a)\xi \otimes \rho_t(b)\xi_t\|^2, \end{aligned}$$

where $\varphi_t^b(x) := \varphi_t(b^*xb)$. Moreover, by the antipode property (1.6.2) we have

$$\begin{aligned} UU^*(\lambda(a)\xi \otimes \rho_t(b)\xi_t) &= U(\lambda(a_{(1)})\xi \otimes \rho_t(S(a_{(2)})b)\xi_t) \\ &= \lambda(a_{(1)})\xi \otimes \rho_t(a_{(2)}S(a_{(3)})b)\xi_t \\ &= \lambda(a_{(1)}\varepsilon(a_{(2)}))\xi \otimes \rho_t(b)\xi_t \\ &= \lambda(a)\xi \otimes \rho_t(b)\xi_t, \end{aligned}$$

which implies that U is an isometry with dense image and therefore extends to a unique unitary operator denoted again by U.

Now the fact that the Markov semigroup $(T_t)_t$ is bounded on $C_r(\mathbb{G})$, that is,

$$\|T_t(a)\|_r = \|\lambda(T_t(a))\|_{B(\mathcal{H})} \leq \|\lambda(a)\|_{B(\mathcal{H})} = \|a\|_r,$$

follows immediately from the relation

$$\lambda(T_t(a)) = E_t(U(\lambda(a) \otimes \mathrm{id}_{\mathcal{H}_t})U^*), \tag{1.6.10}$$

since

$$\begin{aligned} \|\lambda(T_t(a))\| &= \|E_t(U(\lambda(a) \otimes \mathrm{id}_{\mathcal{H}_t})U^*)\| \leq \|U(\lambda(a) \otimes \mathrm{id}_{\mathcal{H}_t})U^*\| \\ &= \|\lambda(a) \otimes \mathrm{id}_{\mathcal{H}_t}\| = \|\lambda(a)\|. \end{aligned}$$

To see that (1.6.10) holds, let us fix $v \in \mathcal{H}$ and $b \in \mathrm{Pol}(\mathbb{G})$ such that $v = \lambda(b)\xi$. Then,

$$\begin{aligned} & E_t(U(\lambda(a) \otimes \mathrm{id}_{\mathcal{H}_t})U^*)v \\ =\ & (\pi_t \circ U \circ (\lambda(a) \otimes \mathrm{id}_{\mathcal{H}_t}) \circ U^* \circ i_t)(\lambda(b)\xi) \\ =\ & (\pi_t \circ U \circ (\lambda(a) \otimes \mathrm{id}_{\mathcal{H}_t}) \circ U^*)\,(\lambda(b)\xi \otimes \xi_t) \\ =\ & \pi_t \circ U \circ (\lambda(a) \otimes \mathrm{id}_{\mathcal{H}_t}) \left(\lambda(b_{(1)})\xi \otimes \rho_t(S(b_{(2)}))\xi_t\right) \\ =\ & \pi_t \circ U \left(\lambda(ab_{(1)})\xi \otimes \rho_t(S(b_{(2)}))\xi_t\right) \\ =\ & \pi_t \left(\lambda(a_{(1)}b_{(1)})\xi \otimes \rho_t(a_{(2)}b_{(2)}S(b_{(3)}))\xi_t\right) \\ =\ & \pi_t \left(\lambda(a_{(1)}b)\xi \otimes \rho_t(a_{(2)})\xi_t\right) \\ =\ & \langle \xi_t, \rho_t(a_{(2)})\xi_t\rangle\, \lambda(a_{(1)}b)\xi \\ =\ & \lambda(a_{(1)}\varphi_t(a_{(2)}))\lambda(b)\xi = \lambda(T_t(a))v. \end{aligned}$$

This way we showed that each T_t extends to a contraction on $C_r(\mathbb{G})$. The extensions form again a semigroup and since both Δ and φ_t are completely positive, T_t is completely positive and contractive. Let us now check that $(T_t)_t$ forms a strongly continuous semigroup on $C_r(\mathbb{G})$.

For a given $a \in C_r(\mathbb{G})$ we choose by density $b \in \mathrm{Pol}(\mathbb{G})$ such that $\|a - b\|_r < \epsilon$. Recall that for $b \in \mathrm{Pol}(\mathbb{G})$, $T_t(b) = \varphi_t \star b = (\mathrm{id} \otimes \varphi_t) \circ \Delta(b)$, where $(\varphi_t)_t$ is the convolution semigroup of states on $\mathrm{Pol}(\mathbb{G})$ (*cf.* Subsection 1.5.1). Then

$$\begin{aligned}
\|T_t(a) - a\|_r &\leq \|T_t(a) - T_t(b)\|_r + \|T_t(b) - b\|_r + \|b - a\|_r \\
&\leq 2\|a - b\|_r + \|(\varphi_t \star b) - b\|_r \\
&\leq 2\epsilon + \sum \|b_{(1)}\varphi_t(b_{(2)}) - b_{(1)}\varepsilon(b_{(2)})\|_r \\
&= 2\epsilon + \sum |\varphi_t(b_{(2)}) - \varepsilon(b_{(2)})| \|b_{(1)}\|_r.
\end{aligned}$$

Since $\lim_{t \to 0+} \varphi_t(b) = \varepsilon(b)$ for any $b \in \mathrm{Pol}(\mathbb{G})$ and the sum is finite, we conclude that

$$\lim_{t \to 0+} \|T_t(a) - a\|_r = 0 \quad \text{for each } a \in C_r(\mathbb{G}). \qquad \square$$

The next results give the characterization of Markov semigroups that are related to Lévy processes on compact quantum groups.

Lemma 1.6.5 *Let* $\mathbb{G} = (\mathsf{A}, \Delta)$ *a compact quantum group and* $T : \mathsf{A} \to \mathsf{A}$ *a completely bounded 2-positive linear map.*

If T *is translation invariant, that is, satisfies*

$$\Delta \circ T = (\mathrm{id} \otimes T) \circ \Delta$$

then T *leaves the null space of the Haar state invariant and induces a translation invariant map* $\tilde{T} : C_r(\mathbb{G}) \to C_r(\mathbb{G})$.

Furthermore, we have $\tilde{T}(V_s) \subseteq V_s$ *for all* $s \in \mathcal{I}$ *and therefore* $\tilde{T}$ *leaves the* $*$*-Hopf algebra* $\mathcal{A} = \mathrm{Pol}(\mathbb{G})$ *invariant.*

2-Positivity means that the map $T_2 : M_2(\mathsf{A}) \to M_2(\mathsf{A})$ defined by

$$T_2\big((a_{jk})_{1 \leq j,k \leq 2}\big) = \big(T(a_{jk})\big)_{1 \leq j,k \leq 2}$$

preserves positivity; this implies that

$$T(b)^* T(b) \leq \|T(\mathbf{1})\| T(b^* b)$$

for all $b \in \mathsf{A}$, *cf.* [Pau02, Exercise 3.4]. Complete boundedness guarantees that $\mathrm{id} \otimes T$ is well defined, *cf.* [Pau02].

Proof Let $b \in N_h = \{a \in \mathsf{A}; h(a^*a) = 0\}$. We can assume that $T \neq 0$, then we have $T(\mathbf{1}) \neq 0$ and

$$\begin{aligned} 0 &\leq \frac{h\big(T(b)^*T(b)\big)}{\|T(\mathbf{1})\|} \leq h\big(T(b^*b)\big) = (h \star h)\big(T(b^*b)\big) \\ &= (h \otimes h)\big(\Delta \circ T(b^*b)\big) = \big(h \otimes (h \circ T)\big)\big(\Delta(b^*b)\big) \\ &= h(b^*b)h\big(T(\mathbf{1})\big) = 0, \end{aligned}$$

where we used idempotence and left invariance of the Haar state. This proves the first claim.

Let $s, s' \in \mathcal{I}$, $s \neq s'$, and $1 \leq j,k \leq n_s$, $1 \leq p,q \leq n_{s'}$. Using again the fact that the Haar state is idempotent, we have

$$\begin{aligned} & h\left(\left(u_{pq}^{(s')}\right)^* \tilde{T}\left(u_{jk}^{(s)}\right)\right) = (h \star h)\left(\left(u_{pq}^{(s')}\right)^* \tilde{T}\left(u_{jk}^{(s)}\right)\right) \\ &= \sum_{r=1}^{n_{s'}} (h \otimes h)\left(\left(\left(u_{pr}^{(s')}\right)^* \otimes \left(u_{rq}^{(s')}\right)^*\right) \Delta\left(\tilde{T}\left(u_{jk}^{(s)}\right)\right)\right) \\ &= \sum_{r=1}^{n_{s'}} \sum_{\ell=1}^{n_s} (h \otimes h)\left(\left(u_{pr}^{(s')}\right)^* \otimes \left(u_{rq}^{(s')}\right)^* \left(u_{j\ell}^{(s)} \otimes \tilde{T}\left(u_{\ell k}^{(s)}\right)\right)\right) \\ &= \sum_{\ell=1}^{n_s} \delta_{ss'} \frac{\overline{f_1((u_{jp}^{(s)})^*)}}{D_s} h\left((u_{\ell q}^{(s')})^* \tilde{T}(u_{\ell k}^{(s)})\right), \end{aligned}$$

that is, $h\left((u_{pq}^{(s')})^* \tilde{T}(u_{jk}^{(s)})\right) = 0$ for all $s, s' \in \mathcal{I}$, with $s \neq s'$, and all $1 \leq j,k \leq n_s$, $1 \leq p,q \leq n_{s'}$. Therefore, $\tilde{T}\big(u_{jk}^{(s)}\big) \in V_s$. □

The following Theorem is the main result in this Section. In classical probability, Lévy processes with values, say, in $\mathbb{R}^d$ or in a Lie group, are exactly the Markov processes whose transition probabilities are time- and space-homogeneous. Time homogeneity means that the transition kernels form a semigroup, and in the group-case space homogeneity has to be interpreted as translation invariance. We can now show that the same characterization also holds for Lévy processes on compact quantum groups.

Theorem 1.6.6 *Let (A, Δ) be a compact quantum group and $(T_t)_{t \geq 0}$ a quantum Markov semigroup on (A, Δ).*

Then $(T_t)_{t\geq 0}$ is the quantum Markov semigroup of a (uniquely determined) Lévy process on Pol$(\mathbb{G})$ *if and only if T_t is translation invariant for all $t \geq 0$.*

Proof If $(T_t)_{t\geq 0}$ comes from a Lévy process on Pol$(\mathbb{G})$, then there exists a generating functional ϕ on Pol$(\mathbb{G})$ such that the generator of the semigroup $(T_t)_{t\geq 0}$ is $L_\phi(a) = \phi \star a$ ($a \in$ Pol$(\mathbb{G})$) and $T_t = \exp(-tL_\phi)$ on Pol$(\mathbb{G})$. Then L_ϕ is translation invariant on Pol$(\mathbb{G})$:

$$(\mathrm{id} \otimes L_\phi) \circ \Delta(a) = a_{(1)} \otimes (\phi \star a_{(2)}) = a_{(1)} \otimes a_{(2)}\phi(a_{(3)})$$
$$= \Delta(a_{(1)})\phi(a_{(2)}) = \Delta \circ L_\phi(a).$$

Next, we observe that the powers of a translation invariant operator are again translation invariant: if $(\mathrm{id} \otimes L_\phi) \circ \Delta = \Delta \circ L_\phi$, then, by induction, we have

$$(\mathrm{id} \otimes L_\phi^n) \circ \Delta = (\mathrm{id} \otimes L_\phi) \circ \Delta \circ L_\phi^{n-1} = \Delta \circ L_\phi^n.$$

Therefore, $T_t = \exp(-tL_\phi) = \sum_{n\geq 0} \frac{(-t)^n}{n!} L_\phi^n$ is also translation-invariant on Pol$(\mathbb{G})$ for each $t \geq 0$. By continuity T_t is translation-invariant on A.

Reciprocally, if T_t is translation invariant, then, by the previous Lemma, it induces a linear map $\tilde{T}_t$ on $C_r(\mathbb{G}) = \mathsf{A}/N_h$ that maps Pol$(\mathbb{G})$ to itself, and so $\varphi_t := \varepsilon \circ \tilde{T}_t$ is well defined on Pol$(\mathbb{G})$ (ε is defined on Pol$(\mathbb{G})$, but may not extend to A). From the Markov semigroup properties of $(T_t)_{t\geq 0}$ we deduce that $(\varphi_t)_{t\geq 0}$ is a convolution semigroup of states on Pol$(\mathbb{G})$. The generating functional of this semigroup defines uniquely a Lévy process on Pol$(\mathbb{G})$, and we have

$$T_t(a) = (\mathrm{id} \otimes \varphi_t)(a) + N_h, \tag{1.6.11}$$

that is, the Markov semigroup $(T_t)_{t\geq 0}$ is equal to convolution with states from the convolution semigroup $(\varphi_t)_{t\geq 0}$, modulo elements of the null space of the Haar state. □

Remark 1.6.7 Equation (1.6.11) can be interpreted as saying that the restriction of $(T_t)_{t\geq 0}$ to Pol$(\mathbb{G})$ coincides h-almost everywhere with the semigroup $(L_{\varphi_t})_{t\geq 0}$ of the associated Lévy process. We do not know if we can remove N_h from Equation (1.6.11)

A closely related result for C^*-bialgebras with the counit, satisfying the residual vanishing at infinity condition, was proved by Lindsay and Skalski (*cf.* [LS11, Proposition 3.2]). This covers, for example the case of universal locally compact quantum groups and, as special case, discrete quantum groups and coamenable compact quantum groups. Their proof is more technical than the algebraic argument we presented here.

1.7 Independences and Convolutions in Noncommutative Probability

Until now we only considered tensor independence, which is the natural generalization of the notion of stochastic independence used in classical probability and also corresponds to the notion of independent observables used in quantum mechanics. However, in quantum probability there exist also other notions of independence. In this section we shall study the most prominent examples: freeness, monotone independence, and boolean independence, and the corresponding convolutions for probability measures on $\mathbb{R}$, $\mathbb{R}_+$, and $\mathbb{T}$ derived from them.

The material in this Section is taken from [Fra09].

In Subsection 1.7.1 we recall the necessary pre-requisites about Cauchy-Stieltjes transforms of probability measures on the real line, the unit circle, or the positive half-line.

In Subsection 1.7.2 we summarize the definition of freeness and the formulae for computing free convolutions. See also, for example, [VDN92, Voi00, BNT05, NS06], and the references therein.

Then we state a short Lemma that we use in the following Sections, see Lemma 1.7.8.

In Subsections 1.7.4 and 1.7.5 we study the monotone and boolean convolutions for probability measures on the real line, the positive half-line, and the unit circle.

1.7.1 Nevanlinna theory and Cauchy–Stieltjes transforms

Denote by $\mathbb{C}^+ = \{z \in \mathbb{C}; \operatorname{Im} z > 0\}$ and $\mathbb{C}^- = \{z \in \mathbb{C}; \operatorname{Im} z < 0\}$ the upper and lower half plane. For μ a probability measure on $\mathbb{R}$ and $z \in \mathbb{C}^+$, we define its *Cauchy–Stieltjes transform* G_μ by

$$G_\mu(z) = \int_{\mathbb{R}} \frac{1}{z - x} \mathrm{d}\mu(x)$$

and its *reciprocal Cauchy–Stieltjes transform* F_μ by

$$F_\mu(z) = \frac{1}{G_\mu(z)}.$$

Denote by $\mathcal{F}$ the following class of holomorphic self-maps,

$$\mathcal{F} = \left\{ F : \mathbb{C}^+ \to \mathbb{C}^+; F \text{ holomorphic and } \inf_{z\in\mathbb{C}^+} \frac{\operatorname{Im} F(z)}{\operatorname{Im} z} = 1 \right\}.$$

The map $\mu \mapsto F_\mu$ defines a bijection between the class $\mathcal{M}_1(\mathbb{R})$ of probability measures on $\mathbb{R}$ and $\mathcal{F}$, as follows from the following theorem.

Theorem 1.7.1 *[Maa92] Let $F : \mathbb{C}^+ \to \mathbb{C}^+$ be holomorphic, then the following are equivalent.*

(i) $\inf_{z\in\mathbb{C}^+} \frac{\operatorname{Im} F(z)}{\operatorname{Im} z} = 1$;
(ii) there exists a $\mu \in \mathcal{M}_1(\mathbb{R})$ such that $F = F_\mu$.

Furthermore, μ is uniquely determined by F.

Similarly, for μ a probability measure on the unit circle $\mathbb{T} = \{z \in \mathbb{C}; |z| = 1\}$ or on the positive half-line $\mathbb{R}_+ = \{x \in \mathbb{R}; x \geq 0\}$, we define

$$\psi_\mu(z) = \int \frac{xz}{1 - xz} \mathrm{d}\mu$$

and

$$K_\mu(z) = \frac{\psi_\mu(z)}{1 + \psi_\mu(z)}$$

for $z \in \mathbb{C}\backslash \operatorname{supp} \mu$.

The map $\mu \mapsto K_\mu$ defines bijections between the class $\mathcal{M}_1(\mathbb{T})$ of probability measures on $\mathbb{T}$ and the class

$$\mathcal{S} = \{K : \mathbb{D} \to \mathbb{D}; K \text{ holomorphic and } K(0) = 0\},$$

where $\mathbb{D} = \{z \in \mathbb{C}; |z| < 1\}$, and between the class $\mathcal{M}_1(\mathbb{R}_+)$ of probability measures on $\mathbb{R}_+$ and the class

$$\mathcal{P} = \left\{ K : \mathbb{C}\backslash\mathbb{R}_+ \to \mathbb{C}\backslash\mathbb{R}_+; \begin{array}{c} K \text{ holomorphic} \\ \lim_{t\nearrow 0} K(t) = 0, K(\overline{z}) = \overline{K(z)}, \\ \arg z \leq \arg K(z) \leq \pi \text{ for all } z \in \mathbb{C}^+ \end{array} \right\},$$

cf. [BB05] and the references therein.

In the following, if X is an operator with distribution $\mu = \mathcal{L}(X,\Omega)$ w.r.t. Ω, then we will write G_X, F_X, φ_X or K_X instead of $G_{\mathcal{L}(X,\Omega)}$, $F_{\mathcal{L}(X,\Omega)}$, $\varphi_{\mathcal{L}(X,\Omega)}$, or $K_{\mathcal{L}(X,\Omega)}$ for the transforms of the distribution of X.

1.7.2 Free convolutions

By $\mathbb{A}_k$ we denote the set of alternating k-tuples of 1's and 2's, i.e.

$$\mathbb{A}_k = \big\{(\varepsilon_1,\ldots,\varepsilon_k) \in \{1,2\}^k; \varepsilon_1 \neq \varepsilon_2 \neq \ldots \neq \varepsilon_k\big\}.$$

Definition 1.7.2 [Voi86] Let $\mathcal{A}_1, \mathcal{A}_2 \subseteq \mathcal{B}(H)$ be two $*$-algebras of bounded operators on a Hilbert space and assume $\mathbf{1} \in \mathcal{A}_i$, $i = 1,2$. Let Ω be a unit vector in H and denote by Φ the vector state associated with to Ω. We say that $\mathcal{A}_1$ and $\mathcal{A}_2$ are *free*, if we have

$$\Phi(X_1 \cdots X_k) = 0$$

for all $k \geq 1$, $\varepsilon \in \mathbb{A}_k$, $X_1 \in \mathcal{A}_{\varepsilon_1}, \ldots, X_k \in \mathcal{A}_{\varepsilon_k}$ such that

$$\Phi(X_1) = \cdots = \Phi(X_k) = 0.$$

Two normal operators X and Y are called *free*, if the algebras $\operatorname{alg}(X) = \{h(X); h \in C_b(\mathbb{C})\}$ and $\operatorname{alg}(Y) = \{h(Y); h \in C_b(\mathbb{C})\}$ they generate are free.

Theorem 1.7.3 *[Maa92, CG05, CG06] Let μ and ν be two probability measures on the real line, with reciprocal Cauchy transforms F_μ and F_ν. Then there exist unique functions $Z_1, Z_2 \in \mathcal{F}$ such that*

$$F_\mu\big(Z_1(z)\big) = F_\nu\big(Z_2(z)\big) = Z_1(z) + Z_2(z) - z$$

for all $z \in \mathbb{C}^+$.

The function $F = F_\mu \circ Z_1 = F_\nu \circ Z_2$ also belongs to $\mathcal{F}$ and is therefore the reciprocal Cauchy transform of some probability measure λ. One defines the *additive free convolution* of μ and ν as this unique probability measure and writes $\mu \boxplus \nu = \lambda$. This is justified by the following theorem.

Theorem 1.7.4 *[Maa92, BV93] Let X and Y be two self-adjoint operators on some Hilbert space H that are free w.r.t. some unit vector $\Omega \in H$. If Ω is cyclic, that is, if*

$$\overline{\operatorname{alg}\{h(X), h(Y); h \in C_b(\mathbb{R})\}\Omega} = H.$$

then $X + Y$ is essentially self-adjoint and the distribution w.r.t. Ω of its closure is equal to the additive free convolution of the distributions of X and Y w.r.t. to Ω, i.e.

$$\mathcal{L}(X+Y,\Omega) = \mathcal{L}(X,\Omega) \boxplus \mathcal{L}(Y,\Omega).$$

One can show that the reciprocal Cauchy transform F_μ of a probability measure $\mu \in \mathcal{M}_1(\mathbb{R})$ is invertible in an appropriate domain, and define the so-called *Voiculescu transform* of μ as

$$\phi_\mu(z) = F^{\langle -1\rangle}(z) - z,$$

where $F^{\langle -1\rangle}$ denotes the inverse of F. This transform linearizes the free convolution, we have

$$\phi_{\mu\boxplus\nu} = \phi_\mu + \phi_\nu.$$

as can be checked using the 'parallelogram equation' in Theorem 1.7.3.

Example 1.7.5 Consider the Bernoulli distribution

$$\mu = \frac{1}{2}(\delta_{+1} + \delta_{-1}).$$

Then we compute

$$G_\mu(z) = \frac{z}{z^2-1} \qquad \text{and} \qquad F_\mu(z) = z - \frac{1}{z}.$$

Therefore,

$$F_\mu^{\langle -1\rangle}(z) = \frac{1}{2}\left(z + \sqrt{z^2+4}\right)$$

and

$$\phi_{\mu\boxplus\mu}(z) = 2\phi_\mu(z) = 2\big(F_\mu^{\langle -1\rangle}(z) - z\big)\big) = \sqrt{z^2+4} - z$$

for z with $\mathrm{Im}(z) > 1$. We get

$$F_{\mu\boxplus\mu}(z) = \sqrt{z^2-4}$$

and see that the free convolution of two Bernoulli distributions is an arcsine distribution,

$$\mu \boxplus \mu = \frac{1}{\pi}\frac{1}{\sqrt{4-x^2}}\mathbf{1}_{]-2,2[}(x)\mathrm{d}x.$$

Note that here the free convolution of two discrete measures has a density and no atoms. In many respects the free convolution has a stronger regularizing effect than classical convolution, see also [BV98, BB04, BB05, Bel08].

Let us now look at the analogous results for the multiplicative free convolutions of probability measures on the unit circle and the positive half-line, *cf.* [BV92, CG05, CG06].

Theorem 1.7.6

(i) Let μ and ν be two probability measures on the unit circle with transforms K_μ and K_ν and whose first moments do not vanish, $\int_{\mathbb{T}} x d\mu(x) \neq 0$, $\int_{\mathbb{T}} x d\nu(x) \neq 0$. Then there exist unique functions $Z_1, Z_2 \in \mathcal{S}$ such that

$$K_\mu(Z_1(z)) = K_\nu(Z_2(z)) = \frac{Z_1(z)Z_2(z)}{z}$$

for all $z \in \mathbb{D}\backslash\{0\}$. The multiplicative free convolution $\lambda = \mu \boxtimes \nu$ is defined as the unique probability measure λ with transform $K_\lambda = K_\mu \circ Z_1 = K_\nu \circ Z_2$.

(ii) Let U and V be two unitary operators on some Hilbert space H that are free w.r.t. some unit vector $\Omega \in H$. Then the products UV and VU are also unitary and their distributions w.r.t. to Ω are equal to the free convolution of the distributions of U and V w.r.t. Ω, that is,

$$\mathcal{L}(UV, \Omega) = \mathcal{L}(VU, \Omega) = \mathcal{L}(U, \Omega) \boxtimes \mathcal{L}(V, \Omega).$$

Theorem 1.7.7

(i) Let μ and ν be two probability measures on the positive half-line such that $\mu \neq \delta_0$, $\nu \neq \delta_0$ and denote their transforms by K_μ and K_ν. Then there exist unique functions $Z_1, Z_2 \in \mathcal{P}$ such that

$$K_\mu(Z_1(z)) = K_\nu(Z_2(z)) = \frac{Z_1(z)Z_2(z)}{z}$$

for all $z \in \mathbb{C}\backslash\mathbb{R}_+$. The multiplicative free convolution $\lambda = \mu \boxtimes \nu$ is defined as the unique probability measure λ with transform $K_\lambda = K_\mu \circ Z_1 = K_\nu \circ Z_2$.

(ii) Let X and Y be two positive operators on some Hilbert space H that are free w.r.t. some unit vector $\Omega \in H$. Assume, furthermore, that Ω is cyclic, that is, that

$$\overline{\mathrm{alg}\{h(X),h(Y);h\in C_b(\mathbb{R})\}\Omega}=H.$$

Then the products $\sqrt{X}Y\sqrt{X}$ and $\sqrt{Y}X\sqrt{Y}$ are essentially self-adjoint and positive, and their distributions w.r.t. to Ω are equal to the free convolution of the distributions of X and Y w.r.t. Ω, that is,

$$\mathcal{L}(\sqrt{X}Y\sqrt{X},\Omega)=\mathcal{L}(\sqrt{Y}X\sqrt{Y},\Omega)=\mathcal{L}(X,\Omega)\boxtimes\mathcal{L}(Y,\Omega).$$

1.7.3 A useful Lemma

To deal with unitary equivalence of possibly unbounded normal operators, we will use several times the following lemma.

Lemma 1.7.8 *Let X and X' be possibly unbounded normal operators on Hilbert spaces H and H'. Assume that there exists a unitary operator $U:H\to H'$ such that*

$$Uf(X)=f(X')U \tag{1.7.1}$$

for any bounded continuous function f on $\mathbb{C}$.

Then $UX=X'U$.

To prove this lemma, one can first prove the analogous statement for self-adjoint operators, e.g., using resolvents, and then apply the properties of the decomposition $X=A+iB$ of normal operators as a linear combination of two commuting self-adjoint operators, *cf.* [Ped89, Proposition 5.1.10]. It is actually sufficient to require Condition (1.7.1) for a much smaller class of functions, for example, compactly supported real-valued C^∞-functions.

1.7.4 Monotone convolutions

Definition 1.7.9 [Mur00] Let $\mathcal{A}_1,\mathcal{A}_2\subset\mathcal{B}(H)$ be two $*$-algebras of bounded operators on a Hilbert space H, and let $\Omega\in H$ be a unit vector. We say that $\mathcal{A}_1$ and $\mathcal{A}_2$ are *monotonically independent* w.r.t. Ω, if we have

$$\langle\Omega,X_1X_2\cdots X_k\Omega\rangle=\left\langle\Omega,\prod_{\kappa:\varepsilon_\kappa=1}X_\kappa\Omega\right\rangle\prod_{\kappa:\varepsilon_\kappa=2}\langle\Omega,X_\kappa\Omega\rangle$$

for all $k\in\mathbb{N}$, $\varepsilon\in\mathbb{A}_k$, $X_1\in\mathcal{A}_{\varepsilon_1},\ldots,X_k\in\mathcal{A}_{\varepsilon_k}$.

Remark 1.7.10

(a) Note that this notion depends on the order, that is, if $\mathcal{A}_1$ and $\mathcal{A}_2$ are monotonically independent, then this does *not* imply that $\mathcal{A}_2$ and $\mathcal{A}_1$ are monotonically independent. In fact, if $\mathcal{A}_1$ and $\mathcal{A}_2$ are monotonically independent and $\mathcal{A}_2$ and $\mathcal{A}_1$ are also monotonically independent, and $\Phi(\cdot) = \langle\Omega, \cdot\,\Omega\rangle$ does not vanish on one of the algebras, then restrictions of Φ to $\mathcal{A}_1$ and $\mathcal{A}_2$ have to be homomorphisms. To prove this for the restriction to, e.g., $\mathcal{A}_1$, take an element $Y \in \mathcal{A}_2$ such that $\Phi(Y) \neq 0$, then

$$\Phi(X_1X_2) = \frac{\Phi(X_1YX_2)}{\Phi(Y)} = \Phi(X_1)\Phi(X_2)$$

for all $X_1, X_2 \in \mathcal{A}_1$.

(b) The sub-algebras are not required to be unital. If $\mathcal{A}_1$ contains the identity operator $\mathbf{1}$, then the restriction of $\Phi(\cdot) = \langle\Omega, \cdot\,\Omega\rangle$ to $\mathcal{A}_2$ has to be a homomorphism, since monotone independence implies

$$\langle\Omega, XY\Omega\rangle = \langle\Omega, X\mathbf{1}Y\Omega\rangle = \langle\Omega, X\Omega\rangle\langle\Omega, Y\Omega\rangle$$

for $X, Y \in \mathcal{A}_2$.

(c) In the definition of monotone independence the condition

$$XYZ = \langle\Omega, Y\Omega\rangle XZ$$

for all $X, Z \in \mathcal{A}_1$, $Y \in \mathcal{A}_2$ is often also imposed. If the state vector Ω is cyclic for the algebra generated by $\mathcal{A}_1$ and $\mathcal{A}_2$, then this is automatically satisfied. Let $X_1, X_3, \ldots, Z_1, Z_3, \ldots \in \mathcal{A}_1$ and $Y, X_2, X_4, \ldots, Z_2, Z_4, \ldots \in \mathcal{A}_2$, then

$$\begin{aligned} &\langle X_1 \cdots X_n\Omega, YZ_1 \cdots Z_m\Omega\rangle = \langle\Omega, X_n^* \cdots X_1^*YZ_1 \cdots Z_m\Omega\rangle \\ = \;& \langle\Omega, Y\Omega\rangle \prod_{k \text{ even}} \langle\Omega, X_k^*\Omega\rangle \prod_{\ell \text{ even}} \langle\Omega, Z_\ell\Omega\rangle\langle X_1X_3 \cdots \Omega, Z_1Z_3 \\ &\qquad\qquad \cdots\Omega\rangle \\ = \;& \langle\Omega, Y\Omega\rangle\langle X_1 \cdots X_n\Omega, Z_1 \cdots Z_m\Omega\rangle, \end{aligned}$$

for all $n, m \geq 1$, that is, $X_1^*YZ_1$ and $\langle\Omega, Y\Omega\rangle X_1^*Z_1$ coincide on the subspace generated by $\mathcal{A}_1$ and $\mathcal{A}_2$ from Ω.

Definition 1.7.11 Let X and Y be two normal operators on a Hilbert space H, not necessarily bounded. We say that X and Y are *monotonically independent* w.r.t. Ω, if the $*$-algebras $\mathrm{alg}_0(X) = \{h(X); h \in C_b(\mathbb{C}), h(0) = 0\}$ and $\mathrm{alg}_0(Y) = \{h(Y); h \in C_b(\mathbb{C}), h(0) = 0\}$ are monotonically independent w.r.t. Ω.

Recall that for monotone independence we have to work with non-unital sub-algebras, *cf.* Remark 1.7.10.(b). This is the reason for the Condition $h(0) = 0$ in the definition of the algebras $\mathrm{alg}_0(X)$ and $\mathrm{alg}_0(Y)$.

Let us now introduce the model we shall use for calculations with monotonically independent operators.

Proposition 1.7.12 *Let μ, ν be two probability measures on $\mathbb{C}$ and define normal operators X and Y on $L^2(\mathbb{C} \times \mathbb{C}, \mu \otimes \nu)$ by*

$$\mathrm{Dom}\, X = \left\{\psi \in L^2(\mathbb{C}\times\mathbb{C}, \mu \otimes \nu); \int_{\mathbb{C}} \left| x \int_{\mathbb{C}} \psi(x,y)\mathrm{d}\nu(y) \right|^2 \mathrm{d}\mu(x) < \infty \right\},$$

$$\mathrm{Dom}\, Y = \left\{\psi \in L^2(\mathbb{C}\times\mathbb{C}, \mu \otimes \nu); \int_{\mathbb{C}\times\mathbb{C}} |y\psi(x,y)|^2 \mathrm{d}\mu \otimes \nu(x,y) < \infty \right\},$$

$$\begin{aligned} (X\psi)(x,y) &= x \int_{\mathbb{C}} \psi(x,y')\mathrm{d}\nu(y'), \\ (Y\psi)(x,y) &= y\psi(x,y). \end{aligned}$$

Then $\mathcal{L}(X, \mathbf{1}) = \mu$, $\mathcal{L}(Y, \mathbf{1}) = \nu$, and X and Y are monotonically independent w.r.t. the constant function $\mathbf{1}$.

Proof Denote by P_2 the orthogonal projection on the space of functions in $L^2(\mathbb{C} \times \mathbb{C}, \mu \otimes \nu)$ that do not depend on the second variable, and by M_x multiplication by the first variable, then $X = M_x P_2$. This operator is normal, we have

$$h(X)\psi(x,y) = (h(x) - h(0)) \int_{\mathbb{C}} \psi(x,y)\mathrm{d}\nu(y) + h(0)\psi(x,y)$$

and $\langle \mathbf{1}, h(X)\mathbf{1}\rangle = \int_{\mathbb{C}} h(x)\mathrm{d}\mu(x)$ for all $h \in C_b(\mathbb{C})$, that is, $\mathcal{L}(X, \mathbf{1}) = \mu$. The operator Y is multiplication by the second variable; it is clearly normal. We have

$$h(Y)\psi(x,y) = h(y)\psi(x,y)$$

and $\langle \mathbf{1}, h(Y)\mathbf{1}\rangle = \int_{\mathbb{C}} h(y)\mathrm{d}\nu(y)$ for all $h \in C_b(\mathbb{C})$, that is, $\mathcal{L}(Y,\mathbf{1}) = \nu$.

Let $f_1,\ldots,f_n,g_1,\ldots,g_{n-1} \in C_b(\mathbb{C})$, $f_1(0) = \cdots = f_n(0) = 0$. Then

$$f_n(X)g_{n-1}(Y)\cdots g_1(Y)f_1(X)\mathbf{1} = \left(\prod_{k=1}^{n-1}\int_{\mathbb{C}} g_k(y)\mathrm{d}\nu(y)\right) f_1(X)\cdots f_n(X)$$

and

$$\begin{aligned}
&\langle \mathbf{1}, f_n(X)g_{n-1}(Y)\cdots g_1(Y)f_1(X)\mathbf{1}\rangle \\
&= \prod_{k=1}^{n-1}\int_{\mathbb{C}} g_k(y)\mathrm{d}\nu(y)\int_{\mathbb{C}} f_1(x)\cdots f_n(x)\mathrm{d}\mu(x) \\
&= \prod_{k=1}^{n-1}\langle \mathbf{1}, g_k(Y)\mathbf{1}\rangle\langle \mathbf{1} f_1(X)\cdots f_n(X)\mathbf{1}\rangle,
\end{aligned}$$

that is, the condition for monotone independence is satisfied in this case. Similarly, one checks the expectation of alternating sequences starting or ending with a function of Y, that is, the cases $g_n(Y)f_n(X)\cdots g_1(Y)f_1(X)$, $f_n(X)g_n(Y)\cdots f_1(X)g_1(Y)$, and $g_n(Y)f_{n-1}(X)\cdots f_1(X)g_1(Y)$. □

The following theorem shows that any pair of monotonically independent normal operators can be reduced to this model.

Theorem 1.7.13 *Let X and Y be two normal operators on a Hilbert space H that are monotonically independent with respect to $\Omega \in H$ and let $\mu = \mathcal{L}(X,\Omega)$, $\nu = \mathcal{L}(Y,\Omega)$.*

Then there exists an isometry $W : L^2(\mathbb{C}\times\mathbb{C}, \mu\otimes\nu) \to H$ such that

$$\begin{aligned}
W^*h(X)W\psi(x,y) &= (h(x)-h(0))\int \psi(x,y)\mathrm{d}\nu(y) + h(0)\psi(x,y), \\
W^*h(Y)W\psi(x,y) &= h(y)\psi(x,y) \qquad (1.7.2)
\end{aligned}$$

for $x,y \in \mathbb{C}$, $\psi \in L^2(\mathbb{C}\times\mathbb{C},\mu\otimes\nu) \cong L^2(\sigma_X,\mu)\otimes L^2(\sigma_Y,\nu)$ and $h \in C_b(\mathbb{C})$.

We have $WL^2(\mathbb{C}\times\mathbb{C},\mu\otimes\nu) = \overline{\mathrm{alg}\{h(X),h(Y);h\in C_b(\mathbb{C})\}\Omega}$.

If the vector $\Omega \in H$ is cyclic for the algebra

$$\mathrm{alg}(X,Y) = \mathrm{alg}\{h(X),h(Y);h\in C_b(\mathbb{C})\}$$

generated by X and Y, then W is unitary.

Proof Define W on simple tensors of bounded continuous functions by

$$Wf \otimes g = g(Y)f(X)\Omega$$

for $f, g \in C_b(\mathbb{C})$. It follows from the monotone independence of X and Y that this defines an isomorphism, since

$$\begin{aligned}\langle Wf_1 \otimes g_1, Wf_2 \otimes g_2\rangle &= \langle \Omega, f_1(X)^* g_1(Y)^* g_2(Y) f_2(X)\Omega\rangle \\ &= \langle \Omega, f_1(X)^* f_2(X)\Omega\rangle\langle \Omega, g_1(Y)^* g_2(Y)\Omega\rangle \\ &= \int \overline{f_1(t)} f_2(t) \mathrm{d}\mu(t) \int \overline{g_1(t)} g_2(t) \mathrm{d}\nu(t).\end{aligned}$$

Since $C_b(\mathbb{C}) \otimes C_b(\mathbb{C})$ is dense in $L^2(\mathbb{C} \times \mathbb{C}, \mu \otimes \nu)$, W extends to a unique isomorphism on $L^2(\mathbb{C} \times \mathbb{C}, \mu \otimes \nu)$.

The relations

$$\begin{aligned}&\langle Wf_1 \otimes g_1, h(X)Wf_2 \otimes g_2\rangle = \langle \Omega, f_1(X)^* g_1(Y)^* h(X) g_2(Y) f_2(X)\Omega\rangle \\ &\quad = \langle \Omega, f_1(X)^* \big(h(X) - h(0)\big) f_2(X)\Omega\rangle\langle \Omega, g_1(Y)^*\Omega\rangle\langle \Omega, g_2(Y)\Omega\rangle \\ &\qquad + h(0)\langle \Omega, f_1(X)^* g_1(Y)^* g_2(Y) f_2(X)\Omega\rangle \\ &\quad = \langle \Omega, g_2(Y)\Omega\rangle \, \langle Wf_1 \otimes g_1, W\big((h - h(0)1\big) f_2 \otimes 1\rangle \\ &\qquad + h(0)\langle Wf_1 \otimes g_1, Wf_2 \otimes g_2\rangle \\ &\quad = \left\langle Wf_1 \otimes g_1, W\left(\int g_2(y)\mathrm{d}\nu(y)(h - h(0)1) f_2 \otimes 1 + h(0) f_2 \otimes g_2\right)\right\rangle\end{aligned}$$

and

$$\begin{aligned}\langle Wf_1 \otimes g_1, h(Y)Wf_2 \otimes g_2\rangle &= \langle \Omega, f_1(X)^* g_1(Y)^* h(Y) g_2(Y) f_2(X)\Omega\rangle \\ &= \langle Wf_1 \otimes g_1, Wf_2 \otimes (hg_2)\rangle\end{aligned}$$

show that we have the desired formulae for simple tensors of functions $f_1, f_2, g_1, g_2 \in C_b(\mathbb{C})$. The general case follows by linearity and continuity. Remark 1.7.10(c) implies

$$\begin{aligned}WL^2(\mathbb{C} \times \mathbb{C}, \mu \otimes \nu) &= \overline{\mathrm{span}\,\{g(Y)f(X)\Omega; f, g \in C_b(\mathbb{C})\}} \\ &= \overline{\mathrm{alg}\{h(X), h(Y); h \in C_b(\mathbb{C})\}\Omega}.\end{aligned}$$

If Ω is cyclic, then W is surjective and therefore unitary. □

Remark 1.7.14 It follows that the joint law of two monotonically independent, normal operators is uniquely determined by their marginal distributions, in the sense that the restriction of $\Phi(\cdot) = \langle \Omega, \cdot \Omega \rangle$ to $\mathrm{alg}(X,Y) = \mathrm{alg}\{h(X), h(Y); h \in C_b(\mathbb{C})\}$ is uniquely determined by $\mathcal{L}(X,\Omega)$ and $\mathcal{L}(Y,\Omega)$. But by Lemma 1.7.8, also computations for unbounded functions of X and Y, e.g., concerning the operators $X+Y$ for self-adjoint X and Y, or $\sqrt{X}Y\sqrt{Y}$ for positive X and Y, reduce to the model introduced in Proposition 1.7.12.

Additive monotone convolution on $\mathcal{M}_1(\mathbb{R})$

Definition 1.7.15 [Mur00] Let μ and ν be two probability measures on $\mathbb{R}$ with reciprocal Cauchy transforms F_μ and F_ν. Then we define the *additive monotone convolution* $\lambda = \mu \triangleright \nu$ of μ and ν as the unique probability measure on $\mathbb{R}$ with reciprocal Cauchy transform $F_\lambda = F_\mu \circ F_\nu$.

It follows from Subsection 1.7.1 that the additive monotone convolution is well defined. Let us first recall some basic properties of the additive monotone convolution.

Proposition 1.7.16 *[Mur00] The additive monotone convolution is associative and $*$-weakly continuous in both arguments. It is affine in the first argument and convolution from the right by a Dirac measure corresponds to translation, that is, $\mu \triangleright \delta_x = T_x^{-1}\mu$ for $x \in \mathbb{R}$, where $T_x : \mathbb{R} \to \mathbb{R}$ is defined by $T_x(t) = t + x$.*

This convolution is not commutative, that is, in general $\mu \triangleright \nu \neq \nu \triangleright \mu$.

Let $x \in \mathbb{R}$ and $0 \leq p \leq 1$. Then one can compute, for example,

$$\delta_x \triangleright \big(p\delta_1 + (1-p)\delta_{-1}\big) = q\delta_{z_1} + (1-q)\delta_{z_2}$$

where

$$\begin{aligned} z_1 &= \frac{1}{2}\left(x + \sqrt{x^2 + 4(2p-1)x + 4}\right), \\ z_2 &= \frac{1}{2}\left(x - \sqrt{x^2 + 4(2p-1)x + 4}\right), \\ q &= \frac{x + 4p - 2 + \sqrt{x^2 + 4(2p-1)x + 4}}{2\sqrt{x^2 + 4(2p-1)x + 4}}. \end{aligned}$$

This example shows that convolution from the left by a Dirac mass is in general not equal to a translation and that the additive monotone convolution is not affine in the second argument.

Note that the continuity and the fact that the monotone convolution is affine in the first argument imply the following formula

$$\mu \rhd \nu = \int_{\mathbb{R}} \delta_x \rhd \nu \, \mathrm{d}\mu(x) \tag{1.7.3}$$

for all $\mu, \nu \in \mathcal{M}_1(\mathbb{R})$.

The following proposition is the key to treat the additive monotone convolution for general probability measures on $\mathbb{R}$.

Proposition 1.7.17 *Let μ and ν be two probability measures on $\mathbb{R}$ and be denoted by M_x and M_y the self-adjoint operators on $L^2(\mathbb{R} \times \mathbb{R}, \mu \otimes \nu)$ defined by multiplication with the coordinate functions. Denote by P_2 the orthogonal projection onto the subspace of functions that do not depend on the second coordinate, that is, $P_2 : L^2(\mathbb{R} \times \mathbb{R}, \mu \otimes \nu) \ni \psi \mapsto \int_{\mathbb{R}} \psi(\cdot, y)\mathrm{d}\nu(y) \in L^2(\mathbb{R} \times \mathbb{R}, \mu \otimes \nu)$. Then $M_x P_2 = P_2 M_x$ and M_y are self-adjoint and monotonically independent w.r.t. the constant function, and the operator $z - M_x P_2 - M_y$ has a bounded inverse for all $z \in \mathbb{C} \backslash \mathbb{R}$, given by*

$$\left((z - M_x P_2 - M_y)^{-1}\psi\right)(x, y) = \frac{\psi(x, y)}{z - y} + \frac{x \int_{\mathbb{R}} \frac{\psi(x, y')}{z - y'} \mathrm{d}\nu(y')}{(z - y)(1 - x G_\nu(z))}. \tag{1.7.4}$$

Proof $M_x P_2$ and M_y are monotonically independent by Proposition 1.7.12.

The first term on the right-hand-side of Equation (1.7.4) is obtained from ψ by multiplication with a bounded function, the second by composition of multiplications with bounded functions and the projection P_2. Equation (1.7.4) therefore clearly defines a bounded operator. To check that it is indeed the inverse of $z - M_x P_2 - M_y$ is straightforward. If we apply the operator $(z - M_x P_2 - M_y)$ to the function

$$\left(\frac{\psi(x, y)}{z - y} + \frac{x \int_{\mathbb{R}} \frac{\psi(x, y')}{z - y'} \mathrm{d}\nu(y')}{(z - y)(1 - x G_\nu(z))}\right),$$

then we can check that this equals

$$\psi(x,y) + \frac{x\int \frac{\psi(x,y')}{z-y'}\mathrm{d}\nu(y')}{1-xG_\nu(z)} - x\int_{\mathbb{R}} \frac{\psi(x,y')}{z-y'}\mathrm{d}\nu(y')$$

$$- x\int_{\mathbb{R}} \frac{x\int_{\mathbb{R}} \frac{\psi(x,y'')}{z-y''}\mathrm{d}\nu(y'')}{(z-y')\big(1-xG_\nu(z)\big)}\mathrm{d}\nu(y')$$

$$=\psi(x,y)$$

$$+\frac{\Big((z-y)-(z-y)\big(1-xG_\nu(z)\big)-xG_\nu(z)(z-y)\Big)x\int_{\mathbb{R}}\frac{\psi(x,y')}{z-y'}\mathrm{d}\nu(y')}{(z-y)\big(1-xG_\nu(z)\big)}$$

$$=\psi(x,y),$$

as it should. □

Theorem 1.7.18 *Let X and Y be two self-adjoint operators on a Hilbert space H that are monotonically independent w.r.t. to a unit vector $\Omega \in H$. Assume, furthermore, that Ω is cyclic, that is, that*

$$\overline{\mathrm{alg}\{h(X),h(Y);h\in C_b(\mathbb{R})\}\Omega} = H.$$

Then $X+Y$ is essentially self-adjoint and the distribution w.r.t. Ω of its closure is equal to the additive monotone convolution of the distributions of X and Y w.r.t. to Ω, that is,

$$\mathcal{L}(X+Y,\Omega) = \mathcal{L}(X,\Omega) \triangleright \mathcal{L}(Y,\Omega).$$

Proof Let $\mu = \mathcal{L}(X,\Omega)$, $\nu = \mathcal{L}(Y,\Omega)$.

By Theorem 1.7.13 and Lemma 1.7.8 it is sufficient to consider the case where X and Y are given by Proposition 1.7.12. Proposition 1.7.17 shows that $z - X - Y$ admits a bounded inverse and therefore that Ran $(z - X - Y)$ is dense for $z \in \mathbb{C}\backslash\mathbb{R}$. By [RS80, Theorem VIII.3] this is equivalent to $X + Y$ being essentially self-adjoint.

Using Equation (1.7.4), we can compute the Cauchy transform of the distribution of the closure of $X + Y$. Let $z \in \mathbb{C}^+$, then we have

$$G_{X+Y}(z) = \langle \Omega, (z - X - Y)^{-1}\Omega \rangle = \left\langle \mathbf{1}, (z - M_x P_2 - M_y)^{-1}\mathbf{1} \right\rangle$$

$$= \left\langle \mathbf{1}, \frac{1}{z-y} + \frac{xG_\nu(z)}{(z-y)(1-xG_\nu(z))} \right\rangle = \int_{\mathbb{R}\times\mathbb{R}} \frac{1}{(z-y)(1-xG_\nu(z))} \mathrm{d}\mu \otimes \nu$$

$$= \int_{\mathbb{R}} \frac{G_\nu(z)}{1 - xG_\nu(z)} \mathrm{d}\mu(x) = G_\mu\left(\frac{1}{G_\nu(z)}\right) = G_\mu(F_\nu(z)),$$

or

$$F_{X+Y}(z) = \frac{1}{G_{X+Y}(z)} = \frac{1}{G_\mu(F_\nu(z))} = F_\mu(F_\nu(z)) = F_{\mu \rhd \nu}(z). \quad \square$$

Multiplicative monotone convolution on $\mathcal{M}_1(\mathbb{R}_+)$

Definition 1.7.19 [Ber05a] Let μ and ν be two probability measures on the positive half-line $\mathbb{R}_+$ with transforms K_μ and K_ν. Then the *multiplicative monotone convolution* of μ and ν is defined as the unique probability measure $\lambda = \mu \rhd \nu$ on $\mathbb{R}_+$ with transform $K_\lambda = K_\mu \circ K_\nu$.

It follows from Subsection 1.7.1 that the multiplicative monotone convolution on $\mathcal{M}_1(\mathbb{R}_+)$ is well defined.

Let us first recall some basic properties of the multiplicative monotone convolution.

Proposition 1.7.20 *The multiplicative monotone convolution $\mathcal{M}_1(\mathbb{R}_+)$ is associative and $*$-weakly continuous in both arguments. It is affine in the first argument and convolution from the right by a Dirac measure corresponds to dilation, that is, $\mu \rhd \delta_\alpha = D_\alpha^{-1}\mu$ for $\alpha \in \mathbb{R}_+$, where $D_\alpha : \mathbb{R}_+ \to \mathbb{R}_+$ is defined by $D_\alpha(t) = \alpha t$.*

This convolution is not commutative, that is, in general $\mu \rhd \nu \neq \nu \rhd \mu$. As in the additive case, it is not affine in the second argument, either, and convolution from the left by a Dirac mass is in general not equal to a dilation.

We want to extend [Fra06a, Corollary 4.3] to unbounded positive operators, that is, we want to show that if X and Y are two positive operators such that $X - \mathbf{1}$ and Y are monotonically independent, then the distribution of $\sqrt{X}Y\sqrt{X}$ is equal to the multiplicative monotone convolution of the distributions of X and Y. By Theorem 1.7.13, it is sufficient to do the calculations for the case where X and Y are constructed from multiplication with the coordinate functions and the projection P_2.

Proposition 1.7.21 *Let μ and ν be two probability measures on $\mathbb{R}_+$, $\nu \neq \delta_0$, and let M_y be the self-adjoint operator on $L^2(\mathbb{R}_+ \times \mathbb{R}_+, \mu \otimes \nu)$ defined by multiplication with the coordinate function $(x, y) \mapsto y$. Define S_x on $L^2(\mathbb{R}_+ \times \mathbb{R}_+, \mu \otimes \nu)$ by*

$$(S_x\psi)(x,y) = (x-1)\int_{\mathbb{R}_+} \psi(x,y)\mathrm{d}\nu(y) + \psi(x,y) \tag{1.7.5}$$

on

$$\mathrm{Dom}\, S_x = \left\{\psi \in L^2(\mathbb{R}_+ \times \mathbb{R}_+, \mu \otimes \nu); \int_{\mathbb{R}_+} x\psi(x,y)\mathrm{d}\nu(y) \in L^2(\mathbb{R}_+, \mu)\right\}.$$

Then $S_x - \mathbf{1}$ and M_y are monotonically independent w.r.t. to the constant function, and the operator $z - \sqrt{S_x}M_y\sqrt{S_x}$ has a bounded inverse for all $z \in \mathbb{C}\backslash\mathbb{R}$, given by

$$\left((z - \sqrt{S_x}M_y\sqrt{S_x})^{-1}\psi\right)(x,y) = \frac{\psi(x,y) + g(x)}{z - y} + h(x). \tag{1.7.6}$$

where

$$\begin{aligned} g(x) &= \frac{\sqrt{x} - x}{(1-x)zG_\nu(z) + x}\int_{\mathbb{R}_+} \psi(x,y)\mathrm{d}\nu(y) \\ &\quad + \frac{z(x-1)}{(1-x)zG_\nu(z) + x}\int_{\mathbb{R}_+} \frac{\psi(x,y)}{z-y}\mathrm{d}\nu(y), \end{aligned}$$

$$\begin{aligned} h(x) &= \frac{(\sqrt{x}-1)^2 G_\nu(z)}{(1-x)zG_\nu(z) + x}\int_{\mathbb{R}_+} \psi(x,y)\mathrm{d}\nu(y) \\ &\quad + \frac{\sqrt{x} - x}{(1-x)zG_\nu(z) + x}\int_{\mathbb{R}_+} \frac{\psi(x,y)}{z-y}\mathrm{d}\nu(y). \end{aligned}$$

Proof Fix $z \in \mathbb{C}^+$. Let $x > 0$, then

$$\mathrm{Im}\frac{z}{z-x} = -\frac{x\,\mathrm{Im}\, z}{(\mathrm{Re}\, z - x)^2 + (\mathrm{Im}\, z)^2} < 0,$$

and therefore

$$\mathrm{Im}\, zG_\nu(z) = \mathrm{Im}\int_{\mathbb{R}_+} \frac{z}{z-x}\mathrm{d}\nu(x) < 0.$$

Similarly, we get $\operatorname{Im} zG_\nu(z) > 0$ for $z \in \mathbb{C}^-$. It follows that the functions in front of the integrals in the definitions of g and h are bounded as functions of x, and therefore g and h are square-integrable. Since $\frac{1}{z-y}$ is bounded too, we see that Equation (1.7.6) defines a bounded operator.

Let us now check that the operator defined in (1.7.6) is the inverse of $z - \sqrt{S_x} M_y \sqrt{S_x}$.

Using the notation of the previous subsection, we can write S_x also as $S_x = M_{x-1}P_2 + \mathbf{1} = M_x P_2 + P_2^\perp$, where $P_2^\perp$ is the projection onto the orthogonal complement of the subspace of functions that do not depend on y. Its square root can be written as $\sqrt{S_x} = M_{\sqrt{x}} P_2 + P_2^\perp = M_{\sqrt{x}-1} P_2 + \mathbf{1}$; it acts as

$$\left(\sqrt{S_x}\psi\right)(x,y) = (\sqrt{x} - 1)\int_{\mathbb{R}_+} \psi(x,y)\mathrm{d}\nu(y) + \psi(x,y)$$

on a function $\psi \in \operatorname{Dom}\sqrt{S_x} \subseteq L^2(\mathbb{R}_+ \times \mathbb{R}_+, \mu \otimes \nu)$.

Since h does not depend on y, we have $\sqrt{S_x}h = \sqrt{x}h$. For g we get

$$\begin{aligned} \left(\sqrt{S_x}\frac{g}{z-y}\right)(x) &= (\sqrt{x}-1)\int_{\mathbb{R}_+} \frac{g(x)}{z-y}\mathrm{d}\nu(y) + \frac{g(x)}{z-y} \\ &= \left((\sqrt{x}-1)G_\nu(z) + \frac{1}{z-y}\right)g(x). \end{aligned}$$

Set $\varphi = \frac{\psi+g}{z-y} + h$. Applying $\sqrt{S_x}$ to φ, we get

$$\begin{aligned} &\left(\sqrt{S_x}\varphi\right)(x,y) \\ =&\frac{\psi(x,y)}{z-y} + \frac{\sqrt{x}-x}{(z-y)\big((1-x)zG_\nu(z)+x\big)}\int_{\mathbb{R}_+}\psi(x,y)\mathrm{d}\nu(y) \\ &+ \frac{z(x-1)}{(z-y)\big((1-x)zG_\nu(z)+x\big)}\int_{\mathbb{R}_+}\frac{\psi(x,y)}{z-y}\mathrm{d}\nu(y) \\ =&\frac{\psi(x,y)+g(x)}{z-y}. \end{aligned}$$

From this we get

$$\left(\left(z - \sqrt{S_x}M_y\sqrt{S_x}\right)\varphi\right)(x,y) = \psi(x,y)$$

after some tedious, but straightforward, computation. □

Remark 1.7.22 If $\nu = \delta_0$, then $M_y = 0$ on $L^2(\mathbb{R}_+ \times \mathbb{R}_+, \mu \otimes \nu)$, and therefore $\sqrt{S_x} M_y \sqrt{S_x} = 0$. This is of course a positive operator, and its distribution is δ_0.

Theorem 1.7.23 *Let X and Y be two positive self-adjoint operators on a Hilbert space H such that $X - \mathbf{1}$ and Y are monotonically independent w.r.t. to a unit vector $\Omega \in H$. Assume, furthermore, that Ω is cyclic, that is,*

$$\overline{\mathrm{alg}\{h(X), h(Y); h \in C_b(\mathbb{R}_+)\}\Omega} = H.$$

Then $\sqrt{X}Y\sqrt{X}$ is essentially self-adjoint and the distribution w.r.t. Ω of its closure is equal to the multiplicative monotone convolution of the distributions of X and Y w.r.t. Ω, that is,

$$\mathcal{L}\left(\sqrt{X}Y\sqrt{X}, \Omega\right) = \mathcal{L}(X, \Omega) \triangleright \mathcal{L}(Y, \Omega).$$

Proof Let $\mu = \mathcal{L}(X, \Omega)$, $\nu = \mathcal{L}(Y, \Omega)$.

By Theorem 1.7.13 it is sufficient to consider the case $X = S_x$ and $Y = M_y$. In this case Proposition 1.7.21 shows that $z - \sqrt{X}Y\sqrt{X}$ has a bounded inverse for all $z \in \mathbb{C}\backslash\mathbb{R}$. This implies that $\mathrm{Ran}(z - \sqrt{X}Y\sqrt{X})$ is dense for all $z \in \mathbb{C}\backslash\mathbb{R}$ and that $\sqrt{X}Y\sqrt{X}$ is essentially self-adjoint, *cf.* [RS80, Theorem VIII.3].

Using Equation (1.7.6), we can compute the Cauchy transform of the distribution of the closure of $\sqrt{X}Y\sqrt{X}$. Let $z \in \mathbb{C}^+$, then we have

$$\begin{aligned} G_{\sqrt{X}Y\sqrt{X}}(z) &= \left\langle \Omega, \left(z - \sqrt{X}Y\sqrt{X}\right)^{-1} \Omega \right\rangle \\ &= \left\langle \mathbf{1}, \left(z - \sqrt{S_x}M_y\sqrt{S_x}\right)^{-1} \mathbf{1} \right\rangle \\ &= \left\langle \mathbf{1}, \frac{1+g_1}{z-y} + h_1 \right\rangle \end{aligned}$$

where

$$\begin{aligned} g_1(x) &= \frac{\sqrt{x} - x + (x-1)zG_\nu(x)}{(1-x)zG_\nu(z) + x} = \frac{\sqrt{x}}{(1-x)zG_\nu(z) + x} - 1, \\ h_1(x) &= \frac{(1-\sqrt{x})G_\nu(z)}{(1-x)zG_\nu(z) + x}. \end{aligned}$$

Therefore,

$$\begin{aligned} G_{\sqrt{X}Y\sqrt{X}}(z) &= \int_{\mathbb{R}_+\times\mathbb{R}_+} \left(\frac{1+g_1(x)}{z-y} + h_1(x)\right) \mathrm{d}\mu\otimes\nu(x,y) \\ &= \int_{\mathbb{R}_+} \frac{G_\nu(z)}{(1-x)zG_\nu(z)+x}\mathrm{d}\mu(x) \\ &= \frac{G_\nu(z)}{zG_\nu(z)-1} G_\mu\left(\frac{zG_\nu(z)}{zG_\nu(z)-1}\right). \end{aligned} \tag{1.7.7}$$

Using the relation

$$G_\mu(z) = \frac{1}{z}\left(\psi_\mu\left(\frac{1}{z}\right)+1\right)$$

to replace the Cauchy transforms by the ψ-transforms, this becomes

$$\psi_{\sqrt{X}Y\sqrt{X}}\left(\frac{1}{z}\right) = \psi_\mu\left(\frac{\psi_\nu(1/z)}{\psi_\nu(1/z)+1}\right),$$

or finally

$$K_{\sqrt{X}Y\sqrt{X}}(z) = K_\mu(K_\nu(z)) = K_{\mu\rhd\nu}(z). \qquad \square$$

Multiplicative monotone convolution on $\mathcal{M}_1(\mathbb{T})$

Definition 1.7.24 [Ber05a] Let μ and ν be two probability measures on the unit circle $\mathbb{T}$ with transforms K_μ and K_ν. Then the *multiplicative monotone convolution* of μ and ν is defined as the unique probability measure $\lambda = \mu \rhd \nu$ on $\mathbb{T}$ with transform $K_\lambda = K_\mu \circ K_\nu$.

It follows from Subsection 1.7.1 that the multiplicative monotone convolution on $\mathcal{M}_1(\mathbb{T})$ is well defined.

Let us first recall some basic properties of the multiplicative monotone convolution.

Proposition 1.7.25 *The multiplicative monotone convolution on the unit circle $\mathbb{T}$ is associative and $*$-weakly continuous in both arguments. It is affine in the first argument and convolution from the right by a Dirac measure corresponds to rotation, that is, $\mu \rhd \delta_{e^{i\vartheta}} = R_\vartheta^{-1}\mu$ for $\vartheta \in [0, 2\pi[$, where $R_\vartheta : \mathbb{T} \to \mathbb{T}$ is defined by $R_\vartheta(t) = e^{i\vartheta}t$*

This convolution is not commutative, that is, in general $\mu \rhd \nu \neq \nu \rhd \mu$. As in the additive case, it is not affine in the second argument, either, and convolution from the left by a Dirac mass is in general not equal to a rotation.

Probability measures on the unit circle arise as distributions of unitary operators and they are completely characterized by their moments. Therefore, the following theorem is a straightforward consequence of [Ber05a] (see also [Fra06a, Theorem 4.1 and Corollary 4.2]).

Theorem 1.7.26 *Let U and V be two unitary operators on a Hilbert space H, $\Omega \in H$ a unit vector and assume, furthermore, that $U - \mathbf{1}$ and V are monotonically independent w.r.t. Ω. Then the products UV and VU are also unitary and their distribution w.r.t. Ω is equal to the multiplicative monotone convolution of the distributions of U and V, that is,*

$$\mathcal{L}(UV,\Omega) = \mathcal{L}(VU,\Omega) = \mathcal{L}(U,\Omega) \rhd \mathcal{L}(V,\Omega). \tag{1.7.8}$$

Remark 1.7.27 Note that the order of the convolution product on the right-hand-side of Equation (1.7.8) depends only on the order in which the operators $U - \mathbf{1}$ and $V - \mathbf{1}$ are monotonically independent, but not on the order in which U and V are multiplied.

1.7.5 Boolean convolutions

Definition 1.7.28 [Spe97, SW97] Let $\mathcal{A}_1, \mathcal{A}_2 \subset \mathcal{B}(H)$ be two $*$-algebras of bounded operators on a Hilbert space H, and let $\Omega \in H$ be a unit vector. We say that $\mathcal{A}_1$ and $\mathcal{A}_2$ are *boolean-independent* w.r.t. Ω, if we have

$$\langle \Omega, X_1X_2\cdots X_k\Omega\rangle = \prod_{\kappa=1}^{k}\langle \Omega, X_\kappa\Omega\rangle$$

for all $k \in \mathbb{N}$, $\varepsilon \in \mathbb{A}_k$, $X_1 \in \mathcal{A}_{\varepsilon_1}, \ldots, X_k \in \mathcal{A}_{\varepsilon_k}$.

Remark 1.7.29 The sub-algebras are not required to be unital. If one of them contains the identity operator $\mathbf{1}$, say $\mathcal{A}_1$, then the restriction of $\Phi(\cdot) = \langle \Omega, \cdot\,\Omega\rangle$ to the other algebra, say $\mathcal{A}_2$, has to be a homomorphism, since the boolean independence implies

$$\langle \Omega, XY\Omega\rangle = \langle \Omega, X\mathbf{1}Y\Omega\rangle = \langle \Omega, X\Omega\rangle\langle \Omega, Y\Omega\rangle$$

for $X, Y \in \mathcal{A}_2$.

Definition 1.7.30 Let X and Y be two normal operators on a Hilbert space H, not necessarily bounded. We say that X and Y are boolean-independent, if the $*$-algebras $\mathrm{alg}_0(X) = \{h(X) : h \in C_b(\mathbb{C}), h(0) = 0\}$ and $\mathrm{alg}_0(Y) = \{h(Y) : h \in C_b(\mathbb{C}), h(0) = 0\}$ are *boolean-independent*.

We will start by characterizing up to unitary transformations the general form of two boolean-independent normal operators. Given a measure space $(M, \mathcal{M}, \mu)$, we shall denote by $L^2(M,\mu)_0$ the orthogonal complement of the constant function, that is,

$$L^2(M,\mu)_0 = \left\{\psi \in L^2(M,\mu); \int_M \psi \mathrm{d}\mu = 0\right\}.$$

Proposition 1.7.31 *Let μ, ν be two probability measures on $\mathbb{C}$ and define normal operators N_x and N_y on $\mathcal{H}_{\mu,\nu} = \mathbb{C} \oplus L^2(\mathbb{C},\mu)_0 \oplus L^2(\mathbb{C},\nu)_0$ by*

$$\mathrm{Dom}\, N_x = \left\{\begin{pmatrix} \alpha \\ \psi_1 \\ \psi_2 \end{pmatrix} \in \mathcal{H}_{\mu,\nu}; \int_{\mathbb{C}} \left|x\left(\psi_1(x)+\alpha\right)\right|^2 \mathrm{d}\mu(x) < \infty\right\},$$

$$\mathrm{Dom}\, N_y = \left\{\begin{pmatrix} \alpha \\ \psi_1 \\ \psi_2 \end{pmatrix} \in \mathcal{H}_{\mu,\nu}; \int_{\mathbb{C}} \left|y\left(\psi_2(y)+\alpha\right)\right|^2 \mathrm{d}\nu(y) < \infty\right\},$$

$$N_x \begin{pmatrix} \alpha \\ \psi_1 \\ \psi_2 \end{pmatrix} = \begin{pmatrix} \int_{\mathbb{C}} x\left(\psi_1(x)+\alpha\right)\mathrm{d}\mu(x) \\ x(\psi_1+\alpha) - \int_{\mathbb{C}} x\left(\psi_1(x)+\alpha\right)\mathrm{d}\mu(x) \\ 0 \end{pmatrix},$$

$$N_y \begin{pmatrix} \alpha \\ \psi_1 \\ \psi_2 \end{pmatrix} = \begin{pmatrix} \int_{\mathbb{C}} y\left(\psi_2(y)+\alpha\right)\mathrm{d}\nu(y) \\ 0 \\ y(\psi_2+\alpha) - \int_{\mathbb{C}} y\left(\psi_2(y)+\alpha\right)\mathrm{d}\nu(y) \end{pmatrix}.$$

Then N_x and N_y are boolean-independent w.r.t. the vector $\omega = \begin{pmatrix} 1 \\ 0 \\ 0 \end{pmatrix}$ and we have $\mathcal{L}(N_x,\omega) = \mu$, $\mathcal{L}(N_y,\omega) = \nu$.

Proof Under the identification $\mathbb{C} \oplus L^2(\mathbb{C},\mu)_0 \oplus L^2(\mathbb{C},\nu)_0 \cong L^2(\mathbb{C},\mu) \oplus L^2(\mathbb{C},\nu)_0$, where

$$\begin{pmatrix} \alpha \\ \psi_1 \\ \psi_2 \end{pmatrix} \cong \begin{pmatrix} \psi_1 + \alpha \\ \psi_2 \end{pmatrix},$$

the operator N_x becomes multiplication by the variable x on $L^2(\mathbb{C},\mu)$. It is clearly normal and we have

$$h(N_x) \begin{pmatrix} \alpha \\ \psi_1 \\ \psi_2 \end{pmatrix} = \begin{pmatrix} \int_{\mathbb{C}} h(x)\big(\alpha + \psi_1(x)\big)\mathrm{d}\mu(x) \\ h(\alpha+\psi_1) - \int_{\mathbb{C}} h(x)\big(\alpha + \psi_1(x)\big)\mathrm{d}\mu(x) \\ h(0)\psi_2 \end{pmatrix}$$

and $\langle \omega, h(N_x)\omega\rangle = \int_{\mathbb{C}} h\mathrm{d}\mu$ for all $h \in C_b(\mathbb{C})$, that is, $\mathcal{L}(N_x,\omega) = \mu$. Similarly,

$$h(N_y) \begin{pmatrix} \alpha \\ \psi_1 \\ \psi_2 \end{pmatrix} = \begin{pmatrix} \int_{\mathbb{C}} h(y)\big(\alpha + \psi_2(y)\big)\mathrm{d}\nu(y) \\ h(0)\psi_1 \\ h(\alpha+\psi_2) - \int_{\mathbb{C}} h(y)\big(\alpha + \psi_2(y)\big)\mathrm{d}\nu(y) \end{pmatrix}$$

for all $h \in C_b(\mathbb{C})$, and $\mathcal{L}(N_y,\omega) = \nu$.

Let $f_1,\ldots,f_n,g_1,\ldots,g_n \in C_b(\mathbb{C})$, with $f_1(0) = \cdots = f_n(0) = g_1(0) = \cdots = g_n(0) = 0$. Then $f_n(N_x)g_{n-1}(N_y)\cdots g_1(N_y)f_1(N_x)\omega$

$$= \begin{pmatrix} \prod_{k=1}^{n} \int_{\mathbb{C}} f_k \mathrm{d}\mu \prod_{\ell=1}^{n-1} \int_{\mathbb{C}} g_\ell \mathrm{d}\nu \\ \prod_{k=1}^{n-1} \int_{\mathbb{C}} f_k \mathrm{d}\mu \prod_{\ell=1}^{n-1} \int_{\mathbb{C}} g_\ell \mathrm{d}\nu \left(f_n - \int_{\mathbb{C}} f_n \mathrm{d}\mu\right) \\ 0 \end{pmatrix}$$

and therefore

$$\begin{aligned} &\langle \omega, f_n(N_x)g_{n-1}(N_y)\cdots g_1(N_y)f_1(N_x)\omega\rangle \\ &= \prod_{k=1}^{n} \int_{\mathbb{C}} f_k \,\mathrm{d}\mu \prod_{\ell=1}^{n-1} \int_{\mathbb{C}} g_\ell \,\mathrm{d}\nu \\ &= \prod_{k=1}^{n} \langle \omega, f_k(N_x)\omega\rangle \prod_{\ell=1}^{n-1} \langle \omega, g_\ell(N_y)\omega\rangle \end{aligned}$$

that is, the condition for boolean independence is satisfied in this case. Similarly, one checks the expectation of $g_n(N_y)f_n(N_x)\cdots$

$g_1(N_y)f_1(N_x)$, $f_n(N_x)g_n(N_y)\cdots f_1(N_x)g_1(N_y)$, and $g_n(N_y)f_{n-1}(N_x)\cdots f_1(N_x)g_1(N_y)$. □

We shall now show that any pair of boolean-independent normal operators can be reduced to this model.

Theorem 1.7.32 *Let X and Y be two normal operators on a Hilbert space H that are boolean-independent w.r.t. to $\Omega \in H$ and let $\mu = \mathcal{L}(X, \Omega), \nu = \mathcal{L}(Y,\Omega)$.*

Then there exists an isometry $W : \mathbb{C} \oplus L^2(\mathbb{C},\mu)_0 \oplus L^2(\mathbb{C},\nu)_0 \to H$ such that

$$W^*h(X)W\begin{pmatrix}\alpha\\ \psi_1\\ \psi_2\end{pmatrix} = \begin{pmatrix}\int_{\mathbb{C}} h(x)\big(\alpha+\psi_1(x)\big)\mathrm{d}\mu(x)\\ h(\alpha+\psi_1) - \int_{\mathbb{C}} h(x)\big(\alpha+\psi_1(x)\big)\mathrm{d}\mu(x)\\ h(0)\psi_2\end{pmatrix}, \tag{1.7.9}$$

$$W^*h(Y)W\begin{pmatrix}\alpha\\ \psi_1\\ \psi_2\end{pmatrix} = \begin{pmatrix}\int_{\mathbb{C}} h(y)\big(\alpha+\psi_2(y)\big)\mathrm{d}\nu(y)\\ h(0)\psi_1\\ h(\alpha+\psi_2) - \int_{\mathbb{C}} h(y)\big(\alpha+\psi_2(y)\big)\mathrm{d}\nu(y)\end{pmatrix}$$

for all $h \in C_b(\mathbb{C})$, $\alpha \in \mathbb{C}$, $\psi_1 \in L^2(\mathbb{C},\mu)_0$, $\psi_2 \in L^2(\mathbb{C},\nu)_0$.

We have

$$W\big(\mathbb{C} \oplus L^2(\mathbb{C},\mu)_0 \oplus L^2(\mathbb{C},\nu)_0\big) = \overline{\mathrm{alg}\{h(X),h(Y) : h \in C_b(\mathbb{C})\}\Omega}.$$

If $\Omega \in H$ is cyclic for the algebra

$$\mathrm{alg}(X,Y) = \mathrm{alg}\{h(X),h(Y) : h \in C_b(\mathbb{C})\}$$

generated by X and Y, then W is unitary.

Proof For a probability measure μ on $\mathbb{C}$, let

$$C_b(\mathbb{C})_{\mu,0} = \left\{f \in C_b(\mathbb{C}); \int_{\mathbb{C}} f(z)\mathrm{d}\mu(x) = 0\right\},$$

then $C_b(\mathbb{C})_{\mu,0}$ is dense in $L^2(\mathbb{C},\mu)_0$.

Define $W : \mathbb{C} \oplus C_b(\mathbb{C})_{\mu,0} \oplus C_b(\mathbb{C})_{\nu,0} \to H$ by

$$W\begin{pmatrix}\alpha\\ f\\ g\end{pmatrix} = \big(\alpha + f(X) + g(Y)\big)\Omega.$$

This is an isometry, since

$$
\begin{aligned}
&\left\langle W\begin{pmatrix}\alpha_1\\ f_1\\ g_1\end{pmatrix}, W\begin{pmatrix}\alpha_2\\ f_2\\ g_2\end{pmatrix}\right\rangle\\
&= \big\langle \big(\alpha_1 + f_1(X) + g_1(Y)\big)\Omega, \big(\alpha_2 + f_2(X) + g_2(Y)\big)\Omega\big\rangle\\
&= \overline{\alpha_1}\alpha_2 + \int_{\mathbb{C}} \overline{f_1(x)} f_2(x)\mathrm{d}\mu(x) + \int_{\mathbb{C}} \overline{g_1(y)} g_2(y)\mathrm{d}\mu(y),
\end{aligned}
$$

where the mixed terms all vanish because $\langle\Omega, f_i(X)\Omega\rangle = \langle\Omega, g_i(Y)\Omega\rangle = 0$ for $i = 1,2$. Therefore, W extends in a unique way to an isometry on $\mathbb{C} \oplus L^2(\mathbb{C},\mu)_0 \oplus L^2(\mathbb{C},\nu)_0$

Let now $h \in C_b(\mathbb{C})$, then we get

$$
\begin{aligned}
&\left\langle W\begin{pmatrix}\alpha_1\\ f_1\\ g_1\end{pmatrix}, h(X)W\begin{pmatrix}\alpha_2\\ f_2\\ g_2\end{pmatrix}\right\rangle\\
&= \big\langle \big(\alpha_1 + f_1(X) + g_1(Y)\big)\Omega, \big(h(X) - h(0)\mathbf{1}\big)\big(\alpha_2 + f_2(X) + g_2(Y)\big)\Omega\big\rangle\\
&\quad + h(0)\left\langle W\begin{pmatrix}\alpha_1\\ f_1\\ g_1\end{pmatrix}, W\begin{pmatrix}\alpha_2\\ f_2\\ g_2\end{pmatrix}\right\rangle\\
&= \big\langle \big(\alpha_1 + f_1(X)\big)\Omega, \big(h(X) - h(0)\mathbf{1}\big)\big(\alpha_2 + f_2(X)\big)\Omega\big\rangle\\
&\qquad\qquad + h(0)\left\langle \begin{pmatrix}\alpha_1\\ f_1\\ g_1\end{pmatrix}, \begin{pmatrix}\alpha_2\\ f_2\\ g_2\end{pmatrix}\right\rangle,
\end{aligned}
$$

because the boolean independence and $\langle\Omega, g_i(Y)\Omega\rangle = 0$ imply that all other terms vanish. But since $\langle\Omega, f_i(Y)\Omega\rangle = 0$, this is equal to

$$
\begin{aligned}
&\big\langle \big(\alpha_1 + f_1(X)\big)\Omega, h(X)\big(\alpha_2 + f_2(X)\big)\Omega\big\rangle\\
&\quad + h(0)\left(\left\langle \begin{pmatrix}\alpha_1\\ f_1\\ g_1\end{pmatrix}, \begin{pmatrix}\alpha_2\\ f_2\\ g_2\end{pmatrix}\right\rangle - \overline{\alpha_1}\alpha_2 - \langle f_1, f_2\rangle\right)\\
&= \left\langle \begin{pmatrix}\alpha_1\\ f_1\\ g_1\end{pmatrix}, \begin{pmatrix}\int h(x)\big(f_2(x)+\alpha_2\big)\mathrm{d}\mu(x)\\ h(f_2+\alpha_2) - \int h(x)\big(f_2(x)+\alpha_2\big)\mathrm{d}\mu(x)\\ h(0)g_2\end{pmatrix}\right\rangle.
\end{aligned}
$$

This proves the first formula. The second formula follows by symmetry.

Let $f, g \in C_b(\mathbb{C})$, $f(0) = 0$, and note that

$$\begin{aligned}
&\left\| f(X)g(Y)\Omega - \int_{\mathbb{C}} g \mathrm{d}\nu \, f(X)\Omega \right\|^2 \\
&= \langle \Omega, g(Y)^* |f(X)|^2 g(Y)\Omega \rangle - \int_{\mathbb{C}} g \mathrm{d}\nu \, \langle \Omega, g(Y)^* |f(X)|^2 \Omega \rangle \\
&\quad - \int_{\mathbb{C}} \overline{g} \mathrm{d}\nu \, \langle \Omega, |f(X)|^2 g(Y)\Omega \rangle + \left(\int_{\mathbb{C}} g \mathrm{d}\nu \right)^2 \langle \Omega, |f(X)|^2 \Omega \rangle \\
&= 0,
\end{aligned}$$

that is, $f(X)g(Y)\Omega = \int_{\mathbb{C}} g \mathrm{d}\nu \, f(X)\Omega$. Similarly $f(Y)g(X)\Omega = \int_{\mathbb{C}} g \mathrm{d}\mu \, f(Y)\Omega$ and thus

$$\begin{aligned}
\overline{\mathrm{alg}\{h(X), h(Y) : h \in C_b(\mathbb{C})\}\Omega} &= \overline{\mathrm{span}\,\{\Omega, f(X)\Omega, f(Y)\Omega; f \in C_b(\mathbb{C})\}} \\
&= W\big(\mathbb{C} \oplus L^2(\mathbb{C}, \mu)_0 \oplus L^2(\mathbb{C}, \nu)_0\big).
\end{aligned}$$

If Ω is cyclic, then W is surjective and therefore unitary. □

Remark 1.7.33 As in the monotone case, *cf.* Remark 1.7.14, this theorem shows that the joint law of bounded functions of X and Y is uniquely determined by $\mathcal{L}(X, \Omega)$ and $\mathcal{L}(Y, \Omega)$. Furthermore, the characterization and computation of the law of unbounded functions of X and Y like, e.g., $X + Y$ or $\sqrt{X}Y\sqrt{Y}$, is also reduced to the model introduced in Proposition 1.7.31.

Additive boolean convolution on $\mathcal{M}_1(\mathbb{R})$

Definition 1.7.34 [SW97] Let μ and ν be two probability measures on $\mathbb{R}$ with reciprocal Cauchy transforms F_μ and F_ν. Then we define the *additive boolean convolution* $\lambda = \mu \uplus \nu$ of μ and ν as the unique probability measure λ on $\mathbb{R}$ with reciprocal Cauchy transform given by

$$F_\lambda(z) = F_\mu(z) + F_\nu(z) - z$$

for $z \in \mathbb{C}^+$.

That the additive boolean convolution is well defined follows from Subsection 1.7.1. It is commutative and associative, $*$-weakly continuous, but not affine, *cf.* [SW97].

Proposition 1.7.35 *Let μ and ν be two probabilities on $\mathbb{R}$ and define operators N_x and N_y as in Proposition 1.7.31. Then N_x and N_y are self-adjoint and boolean-independent w.r.t.* $\omega = \begin{pmatrix} 1 \\ 0 \\ 0 \end{pmatrix}$. *Furthermore, the operator $z - N_x - N_y$ has a bounded inverse for all $z \in \mathbb{C}\backslash\mathbb{R}$, given by*

$$(z - N_x - N_y)^{-1} \begin{pmatrix} \alpha \\ \psi_1 \\ \psi_2 \end{pmatrix} = \begin{pmatrix} \beta \\ \frac{\psi_1 + \beta x - c_x}{z-x} \\ \frac{\psi_2+\beta y - c_y}{z-y} \end{pmatrix}, \tag{1.7.10}$$

where

$$\beta = \frac{\alpha G_\mu(z) G_\nu(z) + G_\nu(z) \int_{\mathbb{R}} \frac{\psi_1(x)}{z-x} \mathrm{d}\mu(x) + G_\mu(z) \int_{\mathbb{R}} \frac{\psi_2(y)}{z-y} \mathrm{d}\nu(y)}{G_\mu(z) + G_\nu(z) - z G_\mu(z) G_\nu(z)}, \tag{1.7.11}$$

and $c_x, c_y \in \mathbb{C}$ have to be chosen such that

$$\int_{\mathbb{R}} \frac{\psi_1(x) + \beta x - c_x}{z - x} \mathrm{d}\mu(x) = 0 = \int_{\mathbb{R}} \frac{\psi_2(y) + \beta y - c_y}{z-y} \mathrm{d}\nu(y). \tag{1.7.12}$$

Note that Equation (1.7.12) yields the following formulae for the constants c_x, c_y,

$$\begin{aligned} c_x &= \frac{\int \frac{\psi_1(x)}{z-x} \mathrm{d}\mu(x) + \beta\big(z G_\mu(z) - 1\big)}{G_\mu(z)}, \\ c_y &= \frac{\int \frac{\psi_2(y)}{z-y} \mathrm{d}\nu(y) + \beta\big(z G_\nu(z) - 1\big)}{G_\nu(z)}. \end{aligned}$$

Proof N_x and N_y are boolean-independent by Proposition 1.7.31.

For $z \in \mathbb{C}^+$, we have $\operatorname{Im} F_\mu(z) \geq \operatorname{Im} z > 0$, $\operatorname{Im} F_\nu(z) \geq \operatorname{Im} z > 0$, and therefore

$$\operatorname{Im} \frac{G_\mu(z) + G_\nu(z) - z G_\mu(z) G_\nu(z)}{G_\mu(z) G_\nu(z)} = \operatorname{Im} \big(F_\mu(z) + F_\nu(z) - z\big) > 0.$$

This shows that the denominator of the right-hand side of Equation (1.7.11) cannot vanish for $z \in \mathbb{C}^+$. Since $G_\mu(\overline{z}) = \overline{G_\mu(z)}$, $G_\nu(\overline{z}) =$

$\overline{G_\nu(z)}$, it can not vanish for z with $\operatorname{Im} z < 0$ either. The functions $\frac{1}{z-x}$ and $\frac{x}{z-x}$ are bounded on $\mathbb{R}$ for $z \in \mathbb{C}\backslash\mathbb{R}$, therefore, Equation (1.7.10) defines a bounded operator.

Let

$$\varphi_1 = \frac{\psi_1 + \beta x - c_x}{z-x} \qquad \text{and} \qquad \varphi_2 = \frac{\psi_2 + \beta y - c_y}{z-y},$$

then

$$\begin{aligned}(z - N_x - N_y)\begin{pmatrix}\beta \\ \varphi_1 \\ \varphi_2\end{pmatrix} &= \begin{pmatrix} z\beta + d_x + d_y \\ (z-x)\varphi_1 - \beta x - d_x \\ (z-y)\varphi_2 - \beta y - d_y\end{pmatrix} \\ &= \begin{pmatrix} z\beta + d_x + d_y \\ \psi_1 - c_x - d_x \\ \psi_2 - c_y - d_y\end{pmatrix}\end{aligned}$$

where

$$d_x = \int x(\varphi_1(x) + \beta)\mathrm{d}\mu(x), \qquad d_y = \int y(\varphi_2(y) + \beta)\mathrm{d}\nu(y).$$

Since $\psi_1 \in L^2(\mathbb{R}, \mu)_0$, $\psi_2 \in L^2(\mathbb{R}, \nu)_0$, integrating over the second and third component in the formula above gives $c_x = -d_x$ and $c_y = -d_y$. Therefore,

$$(z - N_x - N_y)\begin{pmatrix}\beta \\ \varphi_1 \\ \varphi_2\end{pmatrix} = \begin{pmatrix} z\beta - c_x - c_y \\ \psi_1 \\ \psi_2\end{pmatrix}$$

We have to show that the first component is equal to α. We get

$$\begin{aligned} z\beta - c_x - c_y = z\beta &- \frac{\int \frac{\psi_1(x)}{z-x}\mathrm{d}\mu(x) + \beta(zG_\mu(z) - 1)}{G_\mu(z)} \\ &- \frac{\int \frac{\psi_1(x)}{z-x}\mathrm{d}\mu(x) + \beta(zG_\nu(z) - 1)}{G_\nu(z)} \\ = \beta\frac{G_\mu(z) + G_\nu(z) - zG_\mu(z)G_\nu(z)}{G_\mu(z)G_\nu(z)} &- \frac{1}{G_\mu(z)}\int\frac{\psi_1(x)}{z-x}\mathrm{d}\mu(x) \\ &- \frac{1}{G_\nu(z)}\int\frac{\psi_2(y)}{z-y}\mathrm{d}\nu(y)\end{aligned}$$

Substituting Equation (1.7.11) into this expression, we get the desired result $z\beta - c_x - c_y = \alpha$. □

Theorem 1.7.36 *Let X and Y be two self-adjoint operators on a Hilbert space H that are boolean-independent w.r.t. a unit vector $\Omega \in H$ and assume that Ω is cyclic, that is, that*

$$\overline{\mathrm{alg}\{h(X), h(Y); h \in C_b(\mathbb{R})\}\Omega} = H.$$

Then $X + Y$ is essentially self-adjoint and the distribution w.r.t. Ω of the closure of $X + Y$ is equal to the boolean convolution of the distributions of X and Y w.r.t. Ω, that is,

$$\mathcal{L}(X + Y, \Omega) = \mathcal{L}(X, \Omega) \uplus \mathcal{L}(Y, \Omega).$$

Proof Let $\mu = \mathcal{L}(X, \Omega)$, $\nu = \mathcal{L}(Y, \Omega)$.

By Theorem 1.7.32 and Lemma 1.7.8 it is sufficient to consider the case where X and Y are defined as in Proposition 1.7.31. Then Proposition 1.7.35 shows that $z - X - Y$ admits a bounded inverse for all $z \in \mathbb{C}\backslash\mathbb{R}$ and therefore that $\mathrm{Ran}\,(z - X - Y)$ is dense. By [RS80, Theorem VIII.3] this is equivalent to $X + Y$ being essentially self-adjoint.

Using Equation (1.7.10), we can compute the Cauchy transform of the distribution of the closure of $X + Y$. Let $z \in \mathbb{C}^+$, then

$$\begin{aligned} & G_{X+Y}(z) \\ = \;& \langle \Omega, (z - X - Y)^{-1}\Omega \rangle = \left\langle \omega, (z - N_x - N_y)^{-1}\omega \right\rangle \\ = \;& \left\langle \begin{pmatrix} 1 \\ 0 \\ 0 \end{pmatrix}, \frac{G_\mu(z)G_\nu(z)}{G_\mu(z) + G_\nu(z) - zG_\mu(z)G_\nu(z)} \begin{pmatrix} 1 \\ \frac{x - \frac{zG_\mu(z)-1}{G_\mu(z)}}{z-x} \\ \frac{y - \frac{zG_\nu(z)-1}{G_\nu(z)}}{z-y} \end{pmatrix} \right\rangle \\ = \;& \frac{G_\mu(z)G_\nu(z)}{G_\mu(z) + G_\nu(z) - zG_\mu(z)G_\nu(z)}. \end{aligned}$$

Replacing all Cauchy transforms by their reciprocals, this becomes

$$F_{X+Y}(z) = F_\mu(z) + F_\nu(z) - z = F_{\mu \uplus \nu}(z).$$ □

Multiplicative boolean convolution on $\mathcal{M}_1(\mathbb{R}_+)$

Bercovici defined a boolean convolution for probability measures in the positive half-line, *cf.* [Ber06].

Definition 1.7.37 [Ber06] Let μ and ν be two probability measures on $\mathbb{R}_+$ with transforms K_μ and K_ν. If the holomorphic function defined by

$$K(z) = \frac{K_\mu(z)K_\nu(z)}{z} \tag{1.7.13}$$

for $z \in \mathbb{C}\backslash\mathbb{R}_+$ belongs to the class $\mathcal{P}$ introduced in Subsection 1.7.1, then the *multiplicative boolean convolution* $\lambda = \mu \mathbin{\uplus\!\!\!\!\times} \nu$ is defined as the unique probability measure λ on $\mathbb{R}_+$ with transform $K_\lambda = K$.

However, in general the function K defined in Equation (1.7.13) does not belong to $\mathcal{P}$ and in that case the convolution of μ and ν is not defined. Bercovici has shown that for any probability measure μ on $\mathbb{R}_+$ not concentrated in one point there exists an $n \in \mathbb{N}$ such that the n-fold convolution product $\mu^{\uplus\!\!\!\!\times\, n}$ of μ with itself is not defined, *cf.* [Ber06, Proposition 3.1].

This is of course related to the problem that in general the product of two positive operators is not positive. One might hope that taking for example, $\sqrt{X}Y\sqrt{X}$ could lead to a better definition of the multiplicative boolean convolution, since this operator will automatically be positive. This leads to a convolution that is always defined, but that is not associative, *cf.* [Fra09].

Multiplicative boolean convolution on $\mathcal{M}_1(\mathbb{T})$

For completeness we recall the results of [Fra08] for the multiplicative boolean convolution on $\mathcal{M}_1(\mathbb{T})$.

Definition 1.7.38 [Fra08] Let μ and ν be two probability measures on the unit circle $\mathbb{T}$ with transforms K_μ and K_ν. Then the *multiplicative boolean convolution* $\lambda = \mu \mathbin{\uplus\!\!\!\!\times} \nu$ is defined as the unique probability on $\mathbb{T}$ with transform K_λ given by

$$K_\lambda(z) = \frac{K_\mu(z)K_\nu(z)}{z}$$

for $z \in \mathbb{D}$.

It is easy to deduce from Subsection 1.7.1 that the multiplicative boolean convolution on $\mathcal{M}_1(\mathbb{T})$ is well defined. It is associative, commutative, $*$-weakly continuous in both arguments, but not affine.

Theorem 1.7.39 *[Fra08, Theorem 2.2] Let U and V be two unitary operators on a Hilbert space H, $\Omega \in H$ a unit vector and assume, furthermore, that $U - \mathbf{1}$ and $V - \mathbf{1}$ are boolean-independent w.r.t. Ω. Then the products UV and VU are also unitary and their distribution w.r.t. Ω is equal to the multiplicative boolean convolution of the distributions of U and V, that is,*

$$\mathcal{L}(UV,\Omega) = \mathcal{L}(VU,\Omega) = \mathcal{L}(U,\Omega) \uplus\!\!\!\times\, \mathcal{L}(V,\Omega).$$

1.8 The Five Universal Independences

In classical probability theory there exists only one canonical notion of independence. However, in quantum probability many different notions of independence have been used, e.g., to obtain central limit theorems or to develop a quantum stochastic calculus. If one requires that the joint law of two independent random variables should be determined by their marginals, then an independence gives rise to a product of algebraic probability spaces. Imposing certain natural conditions, e.g., that functions of independent random variables should again be independent and that the underlying product is associative, it becomes possible to classify all possible notions of independence. This program has been carried out by Schürmann [Sch95a], Speicher [Spe97], Ben Ghorbal and Schürmann [BGS99][BGS02], and Muraki [Mur03, Mur02]. See also the recent generalization by Gerhold and Lachs [GL14]. If one imposes a normalization condition of the correlations of length two, then there are exactly five universal independences: tensor independence, freeness, boolean independence, and monotone and anti-monotone independence. Without this condition there are five one- or two- parameter families, *cf.* [BGS02, GL14].

In this section we will present the results of these classifications.

Universal independences and universal products are of interest for us, because it turns out that it is possible to define Lévy processes for them. However, since bialgebras are built on the tensor products of algebras, they can only be used for Lévy processes with tensor-independent increments. In order to define and study Lévy processes whose increments are free, monotonically independent, or boolean-independent, bialgebras have to be replaced by dual groups, which we shall introduce in Section 1.9.

These universal products of quantum probability spaces have an interpretation in category theory; they are tensor products equipped with canonical inclusions of the factors into the product. Furthermore, it turns out that boolean, monotone, and anti-monotone independence can be reduced to tensor independence in a similar way as the bosonization of Fermi independence [HP86] or the symmetrization of [Sch93, Section 3], *cf.* [Fra02, Fra03c, Fra06b].

Let us recall the definition of a coproduct, it will be underlying many constructions in this Section.

Definition 1.8.1 Let A and B be two objects in some category. A triple $(A\coprod B, i_A, i_B)$ consisting of an object $A\coprod B$ and morphisms $i_A : A \to A\coprod B$ and $i_B : B \to A\coprod B$ is called a *coproduct* of A and B, if for any triple (D, k_A, k_B) consisting of an object D and morphisms $k_A : A \to D$ and $k_B : B \to D$ there exists a unique morphism $h : A\coprod B \to D$ such that the following diagram commutes

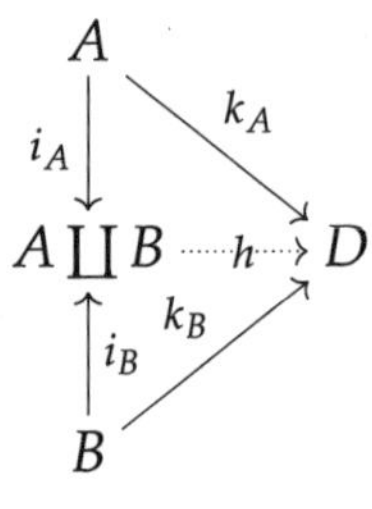

This means that h satisfies $k_A = h \circ i_A$ and $k_B = h \circ i_B$.

The free product of associative algebras is an example of a coproduct, it will play an important role in our discussion.

Example 1.8.2 Denote by $\mathfrak{Alg}$ the category whose objects are unital associative algebras and whose morphisms are unit-preserving algebra homomorphisms.

This category has a coproduct, it is also called *free product* of algebras *with* identification of the units. Let us recall its defining universal property. Let $\{\mathcal{A}_k\}_{k\in I}$ be a family of unital algebras and

$\coprod_{k\in I} \mathcal{A}_k$ their free product, with canonical inclusions $\{i_k : \mathcal{A}_k \to \coprod_{k\in I} \mathcal{A}_k\}_{k\in I}$. If $\mathcal{B}$ is any unital algebra, equipped with unital algebra homomorphisms $\{i'_k : \mathcal{A}_k \to \mathcal{B}\}_{k\in I}$, then there exists a unique unital algebra homomorphism $h : \coprod_{k\in I} \mathcal{A}_k \to \mathcal{B}$ such that

$$h \circ i_k = i'_k, \qquad \text{for all} \quad k \in I.$$

It follows from the universal property that for any pair of unital algebra homomorphisms $j_1 : \mathcal{A}_1 \to \mathcal{B}_1$, $j_2 : \mathcal{A}_2 \to \mathcal{B}_2$, there exists a unique unital algebra homomorphism $j_1 \coprod j_2 : \mathcal{A}_1 \coprod \mathcal{A}_2 \to \mathcal{B}_1 \coprod \mathcal{B}_2$ such that the diagram commutes

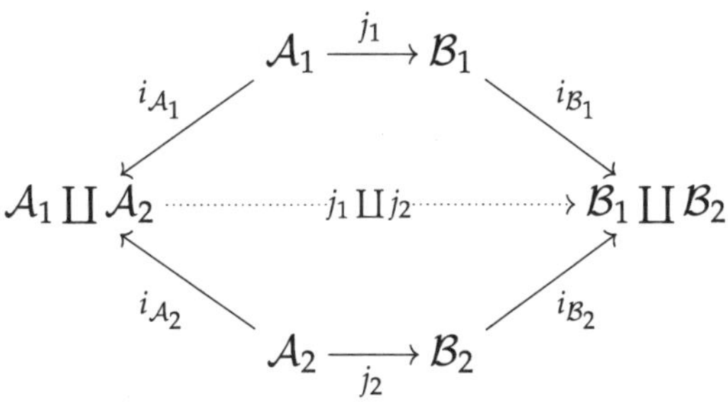

The free product $\coprod_{k\in I} \mathcal{A}_k$ can be constructed as a sum of tensor products of the $\mathcal{A}_k$, where neighboring elements in the product belong to different algebras. For simplicity, we illustrate this only for the case of the free product of two algebras. Let

$$\mathbb{A} = \bigcup_{n\in\mathbb{N}} \{\epsilon \in \{1,2\}^n | \epsilon_1 \neq \epsilon_2 \neq \cdots \neq \epsilon_n\}$$

and decompose $\mathcal{A}_i = \mathbb{C}\mathbf{1} \oplus \mathcal{A}_i^0$, $i = 1, 2$, into a direct sum of vector spaces. As a coproduct $\mathcal{A}_1 \coprod \mathcal{A}_2$ is unique up to isomorphism; therefore, the construction does not depend on the choice of the decompositions.

Then $\mathcal{A}_1 \coprod \mathcal{A}_2$ can be constructed as

$$\mathcal{A}_1 \coprod \mathcal{A}_2 = \bigoplus_{\epsilon\in\mathbb{A}} \mathcal{A}^\epsilon,$$

where $\mathcal{A}^\emptyset = \mathbb{C}$, $\mathcal{A}^\epsilon = \mathcal{A}^0_{\epsilon_1} \otimes \cdots \otimes \mathcal{A}^0_{\epsilon_n}$ for $\epsilon = (\epsilon_1, \ldots, \epsilon_n)$. The multiplication in $\mathcal{A}_1 \coprod \mathcal{A}_2$ is inductively defined by

$$\begin{aligned}(a_1 \otimes \cdots \otimes a_n) \cdot (b_1 \otimes \cdots \otimes b_m) \\ = \begin{cases} a_1 \otimes \cdots \otimes (a_n \cdot b_1) \otimes \cdots \otimes b_m & \text{if } \epsilon_n = \delta_1, \\ a_1 \otimes \cdots \otimes a_n \otimes b_1 \otimes \cdots \otimes b_m & \text{if } \epsilon_n \neq \delta_1, \end{cases}\end{aligned}$$

for $a_1 \otimes \cdots \otimes a_n \in \mathcal{A}^\epsilon$, $b_1 \otimes \cdots \otimes b_m \in \mathcal{A}^\delta$. Note that in the case $\epsilon_n = \delta_1$ the product $a_n \cdot b_1$ is not necessarily in $\mathcal{A}^0_{\epsilon_n}$, but is in general a sum of a multiple of the unit of $\mathcal{A}_{\epsilon_n}$ and an element of $\mathcal{A}^0_{\epsilon_n}$. We have to identify $a_1 \otimes \cdots a_{n-1} \otimes 1 \otimes b_2 \otimes \cdots b_m$ with $a_1 \otimes \cdots \otimes a_{n-1} \cdot b_2 \otimes \cdots b_m$.

Since $\coprod$ is the coproduct of a category, it is commutative and associative in the sense that there exist natural isomorphisms

$$\begin{aligned} \gamma_{\mathcal{A}_1,\mathcal{A}_2} &: \mathcal{A}_1 \coprod \mathcal{A}_2 \overset{\cong}{\to} \mathcal{A}_2 \coprod \mathcal{A}_1, \\ \alpha_{\mathcal{A}_1,\mathcal{A}_2,\mathcal{A}_3} &: \mathcal{A}_1 \coprod (\mathcal{A}_2 \coprod \mathcal{A}_3) \overset{\cong}{\to} (\mathcal{A}_1 \coprod \mathcal{A}_2) \coprod \mathcal{A}_3 \end{aligned} \tag{1.8.1}$$

for all unital algebras $\mathcal{A}_1$, $\mathcal{A}_2$, $\mathcal{A}_3$. Let $i_\ell : \mathcal{A}_\ell \to \mathcal{A}_1 \coprod \mathcal{A}_2$ and $i'_\ell : \mathcal{A}_\ell \to \mathcal{A}_2 \coprod \mathcal{A}_1$, $\ell = 1, 2$ be the canonical inclusions. The commutativity constraint $\gamma_{\mathcal{A}_1,\mathcal{A}_2} : \mathcal{A}_1 \coprod \mathcal{A}_2 \to \mathcal{A}_2 \coprod \mathcal{A}_1$ maps an element of $\mathcal{A}_1 \coprod \mathcal{A}_2$ of the form $i_1(a_1)i_2(b_1) \cdots i_2(b_n)$ with $a_1, \ldots, a_n \in \mathcal{A}_1$, $b_1, \ldots, b_n \in \mathcal{A}_2$ to

$$\gamma_{\mathcal{A}_1,\mathcal{A}_2}\big(i_1(a_1)i_2(b_1) \cdots i_2(b_n)\big) = i'_1(a_1)i'_2(b_1) \cdots i'_2(b_n) \in \mathcal{A}_2 \coprod \mathcal{A}_1.$$

Exercise 1.8.3 We also consider non-unital algebras. Show that the *free product* of algebras *without* identification of units is a coproduct in the category $\mathfrak{NuAlg}$ of non-unital (or rather not necessarily unital) algebras. Give an explicit construction for the free product of two non-unital algebras.

Exercise 1.8.4 If we want to have a notion of positivity, we have to work with involutive algebras. Define the categories $*$-$\mathfrak{Alg}$ and $*$-$\mathfrak{NuAlg}$ of involutive unital or not necessarily unital algebras as above, but requiring in addition that the morphisms respect also the involution, that is, that $j(a^*) = j(a)^*$ for all a in the domain of j. Show that these categories also have coproducts, which can be constructed essentially as in the non-involutive case.

Exercise 1.8.5 Show that one can define a functor from the category of non-unital algebras $\mathfrak{NuAlg}$ to the category of unital algebras $\mathfrak{Alg}$. For an algebra $\mathcal{A} \in \mathrm{Ob}\,\mathfrak{NuAlg}$, $\tilde{\mathcal{A}}$, define $\tilde{\mathcal{A}} = \mathbb{C}\mathbf{1} \oplus \mathcal{A}$ as a vector space and equip this with the multiplication

$$(\lambda \mathbf{1} + a)(\lambda' \mathbf{1} + a') = \lambda\lambda' \mathbf{1} + \lambda' a + \lambda a' + aa'$$

for $\lambda, \lambda' \in \mathbb{C}$, $a, a' \in \mathcal{A}$. We will call $\tilde{\mathcal{A}}$ the *unitization* of $\mathcal{A}$. Note that $\mathcal{A} \cong 0\mathbf{1} + \mathcal{A} \subseteq \tilde{\mathcal{A}}$ is not only a subalgebra, but even an ideal in $\tilde{\mathcal{A}}$.

How is this functor defined on the morphisms?

Show that the following relation holds between the free product *with* identification of units $\coprod_{\mathfrak{Alg}}$ and the free product *without* identification of units $\coprod_{\mathfrak{NuAlg}}$,

$$\mathcal{A}_1 \widetilde{\coprod_{\mathfrak{NuAlg}}} \mathcal{A}_2 \cong \tilde{\mathcal{A}}_1 \coprod_{\mathfrak{Alg}} \tilde{\mathcal{A}}_2$$

for all $\mathcal{A}_1, \mathcal{A}_2 \in \operatorname{Ob} \mathfrak{NuAlg}$.

Note, furthermore, that the range of this functor consists of all algebras that admit a decomposition of the form $\mathcal{A} = \mathbb{C}\mathbf{1} \oplus \mathcal{A}_0$, where $\mathcal{A}_0$ is a subalgebra. This is equivalent to having a one-dimensional representation. The functor is not surjective, e.g., the algebra $\mathcal{M}_2$ of 2×2-matrices can not be obtained as a unitization of some other algebra.

1.8.1 Algebraic probability spaces

From the four categories $\mathfrak{Alg}$, $\mathfrak{NuAlg}$, $*$-$\mathfrak{Alg}$, and $*$-$\mathfrak{NuAlg}$, we form the four categories $\mathfrak{AlgProb}$, $\mathfrak{NuAlgProb}$, $*$-$\mathfrak{AlgProb}$, and $*$-$\mathfrak{NuAlgProb}$ of *algebraic probability spaces* with or without unit and with or without involution. The objects in these categories of algebraic probability spaces are pairs (A, ϕ) consisting of an algebra A taken from the corresponding category of algebras, and a linear functional $\phi : A \to \mathbb{C}$. In the two categories $\mathfrak{AlgProb}$ and $*$-$\mathfrak{AlgProb}$ constructed from unital algebras we impose the condition $\phi(\mathbf{1}) = 1$; in the two categories $*$-$\mathfrak{AlgProb}$ and $*$-$\mathfrak{NuAlgProb}$ built from involutive algebras we impose the additional condition

$$\phi(a^*a) \geq 0 \qquad \text{for all } a \in A.$$

The morphisms in the categories $\mathfrak{AlgProb}$, $\mathfrak{NuAlgProb}$, $*$-$\mathfrak{AlgProb}$, and $*$-$\mathfrak{NuAlgProb}$ are those morphisms from the categories $\mathfrak{Alg}$, $\mathfrak{NuAlg}$, $*$-$\mathfrak{Alg}$, and $*$-$\mathfrak{NuAlg}$, respectively, which preserve furthermore the functionals, that is, $j : (A, \phi_A) \to (B, \phi_B)$ has to satisfy the condition

$$\phi_B \circ j = \phi_A.$$

We can also consider the subcategories $\mathfrak{CommAlgProb}$, $\mathfrak{CommNuAlgProb}$, $*$-$\mathfrak{CommAlgProb}$, and $*$-$\mathfrak{CommNuAlgProb}$ of *commutative* algebraic probability spaces, obtained from $\mathfrak{AlgProb}$,

$\mathfrak{NuAlgProb}$, $*$-$\mathfrak{AlgProb}$, and $*$-$\mathfrak{NuAlgProb}$ by restricting to commutative algebras.

It is the category $*$-$\mathfrak{AlgProb}$ that has as objects the $*$-algebraic probility spaces with which we worked in Section 1.5, when we developed the theory of Lévy processes on involutive bialgebras.

The categories of algebraic probability spaces defined above do not have coproducts. However, they have interesting products, as we shall see in this section. Imposing several natural conditions, such as functoriality and associativity, it becomes possible to classify all such products, see Subsection 1.8.4. These so-called universal products of algebraic probability spaces allow to define the universal notions of independence, see Definition 1.9.1.

Definition 1.8.6 Let $\mathfrak{C}$ be any category equiped with a map $\mathcal{P}$ that associates to any pair of objects $A, B \in \mathrm{Ob}(\mathcal{C})$ a triple

$$\mathcal{P}(A,B) = (D, j_A, j_B)$$

consisting of an object D and morphisms $j_A : A \to D$ and $j_B : B \to D$. Then we say that two morphisms $k_A : A \to C$ and $k_B : B \to C$ are *independent w.r.t. to* $\mathcal{P}$ if there exists a morphism $h : D \to C$ such that $k_A = h \circ j_A$ and $k_B = h \circ j_B$, that is, the diagram commutes.

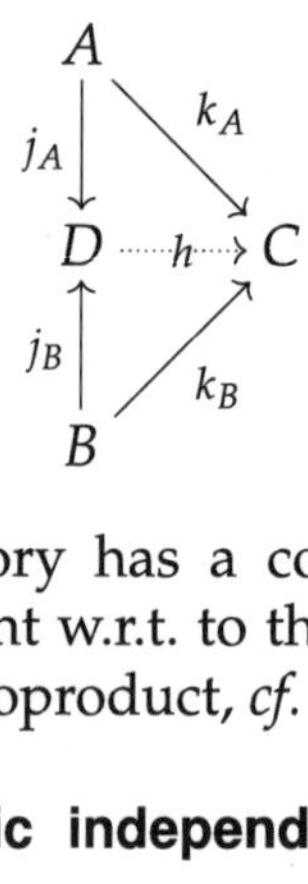

Remark 1.8.7 If a category has a coproduct, then all pairs of morphisms are independent w.r.t. to the coproduct, simply by the universal property of the coproduct, *cf.* Definition 1.8.1.

1.8.2 Classical stochastic independence and the product of probability spaces

Let us first consider the product of classical probability spaces and its relation to classical independence from a category theoretical point of view.

Two random variables $X_1 : (\Omega, \mathcal{F}, P) \to (E_1, \mathcal{E}_1)$ and $X_2 : (\Omega, \mathcal{F}, P) \to (E_2, \mathcal{E}_2)$, defined on the same probability space $(\Omega, \mathcal{F}, P)$

and with values in two possibly distinct measurable spaces $(E_1, \mathcal{E}_1)$ and $(E_2, \mathcal{E}_2)$, are called *stochastically independent* (or simply *independent*) w.r.t. P, if the σ-algebras $X_1^{-1}(\mathcal{E}_1)$ and $X_2^{-1}(\mathcal{E}_2)$ are independent w.r.t. P, that is, if

$$P\big((X_1^{-1}(M_1) \cap X_2^{-1}(M_2)\big) = P\big((X_1^{-1}(M_1)\big)P\big(X_2^{-1}(M_2)\big)$$

holds for all $M_1 \in \mathcal{E}_1$, $M_2 \in \mathcal{E}_2$. If there is no danger of confusion, then the reference to the measure P is often omitted.

This definition can easily be extended to arbitrary families of random variables. A family $\big(X_j : (\Omega, \mathcal{F}, P) \to (E_j, \mathcal{E}_j)\big)_{j \in J}$, indexed by some set J, is called independent, if

$$P\left(\bigcap_{k=1}^{n} X_{j_k}^{-1}(M_{j_k})\right) = \prod_{k=1}^{n} P\big(X_{j_k}^{-1}(M_{j_k})\big)$$

holds for all $n \in \mathbb{N}$ and all choices of indices $k_1, \ldots, k_n \in J$ with $j_k \neq j_\ell$ for $j \neq \ell$, and all choices of measurable sets $M_{j_k} \in \mathcal{E}_{j_k}$.

There are many equivalent formulations for independence; consider, e.g., the following proposition.

Proposition 1.8.8 Let X_1 and X_2 be two real-valued random variables. The following are equivalent.

(i) *X_1 and X_2 are independent.*

(ii) *For all bounded measurable functions f_1, f_2 on $\mathbb{R}$ we have*

$$\mathbb{E}\big(f_1(X_1)f_2(X_2)\big) = \mathbb{E}\big(f_1(X_1)\big)\mathbb{E}\big(f_2(X_2)\big).$$

(iii) *The probability space $(\mathbb{R}^2, \mathcal{B}(\mathbb{R}^2), P_{(X_1,X_2)})$ is the product of the probability spaces $(\mathbb{R}, \mathcal{B}(\mathbb{R}), P_{X_1})$ and $(\mathbb{R}, \mathcal{B}(\mathbb{R}), P_{X_2})$, that is,*

$$P_{(X_1,X_2)} = P_{X_1} \otimes P_{X_2}.$$

We see that stochastic independence can be reinterpreted as a rule to compute the joint distribution of two random variables from their marginal distributions. More precisely, their joint distribution can be computed as a product of their marginal distributions. This product is associative and can also be iterated to compute the joint distribution of more than two independent random variables.

The classifications of independence for noncommutative probability spaces [Spe97, BGS99, BG01, Mur03, Mur02] that we are interested in are based on redefining independence as a product satisfying certain natural axioms.

1.8.3 Products of algebraic probability spaces

We will now define several products on the categories of algebraic probability spaces. In Subsection 1.8.4 we shall see that these products are the only products satisfying the conditions of functoriality and associativity.

Example: Tensor independence in the category of algebraic probability spaces

Let us first consider the usual tensor product on the categories $\mathfrak{AlgProb}$ and $*$-$\mathfrak{AlgProb}$. The product of $(\mathcal{A}, \varphi_1)$ and $\mathcal{A}_2, \varphi_2)$ is $(\mathcal{A}_1 \otimes \mathcal{A}_2, \varphi_1 \otimes \varphi_2)$, where the unital algebra structure of $\mathcal{A}_1 \otimes \mathcal{A}_2$ is defined by

$$\begin{aligned} \mathbf{1}_{\mathcal{A}_1 \otimes \mathcal{A}_2} &= \mathbf{1}_{\mathcal{A}_1} \otimes \mathbf{1}_{\mathcal{A}_2}, \\ (a_1 \otimes a_2)(b_1 \otimes b_2) &= a_1 b_1 \otimes a_2 b_2, \end{aligned}$$

and the new functional is defined by

$$(\varphi_1 \otimes \varphi_2)(a_1 \otimes a_2) = \varphi_1(a_1)\varphi_2(a_2),$$

for all $a_1, b_1 \in \mathcal{A}_1$, $a_2, b_2 \in \mathcal{A}_2$.

Note that there exist canonical morphisms from the objects $(\mathcal{A}, \varphi_1)$ and $(\mathcal{A}_2, \varphi_2)$ to the product $(\mathcal{A}_1 \otimes \mathcal{A}_2, \varphi_1 \otimes \varphi_2)$, defined by

$$\begin{aligned} i_{\mathcal{A}_1}(a_1) &= a_1 \otimes \mathbf{1}_{\mathcal{A}_2}, \\ i_{\mathcal{A}_2}(a_2) &= \mathbf{1}_{\mathcal{A}_1} \otimes a_2, \end{aligned}$$

for $a_1 \in \mathcal{A}_1$, $a_2 \in \mathcal{A}_2$.

If $(\mathcal{A}, \varphi_1)$ and $\mathcal{A}_2, \varphi_2)$ are $*$-algebraic probability spaces, if the underlying algebras have an involution and the functionals are states, then an involution is defined on $\mathcal{A}_1 \otimes \mathcal{A}_2$ by $(a_1 \otimes a_2)^* = a_1^* \otimes a_2^*$ and $\varphi_1 \otimes \varphi_2$ is again a state.

The notion of independence associated to this product with inclusions is the usual notion of *Bose* or *tensor independence* used in quantum probability, e.g., by Hudson and Parthasarathy, see also Definition 1.5.1.

Proposition 1.8.9 *Two quantum random variables $j_1 : (\mathcal{B}_1, \psi_1) \to (\mathcal{A}, \varphi)$ and $j_2 : (\mathcal{B}_2, \psi_2) \to (\mathcal{A}, \varphi)$, defined on algebraic probability spaces $(\mathcal{B}_1, \psi_1), (\mathcal{B}_2, \psi_2)$ and with values in the same algebraic probability space $(\mathcal{A}, \varphi)$ are independent if and only if the following two conditions are satisfied.*

(i) The images of j_1 and j_2 commute, that is,

$$\left[j_1(a_1), j_2(a_2)\right] = 0,$$

for all $a_1 \in \mathcal{A}_1$, $a_2 \in \mathcal{A}_2$.

(ii) φ satisfies the factorization property

$$\varphi\big(j_1(a_1)j_2(a_2)\big) = \varphi\big(j_1(a_1)\big)\varphi\big(j_2(a_2)\big),$$

for all $a_1 \in \mathcal{A}_1$, $a_2 \in \mathcal{A}_2$.

Example 1.8.10 (The universal version of the tensor product)
We can also use the tensor product to define a product

$$\varphi_1 \tilde{\otimes} \varphi_2 : \mathcal{A}_1 \coprod \mathcal{A}_2 \to \mathbb{C}$$

for two functionals $\varphi_i : \mathcal{A}_i \to \mathbb{C}$, simply by composing the tensor product $\varphi_1 \otimes \varphi_2 : \mathcal{A}_1 \otimes \mathcal{A}_2 \to \mathbb{C} \otimes \mathbb{C} \cong \mathbb{C}$ with the unique morphism $R : \mathcal{A}_1 \coprod \mathcal{A}_2 \to \mathcal{A}_1 \otimes \mathcal{A}_2$ that makes the diagram

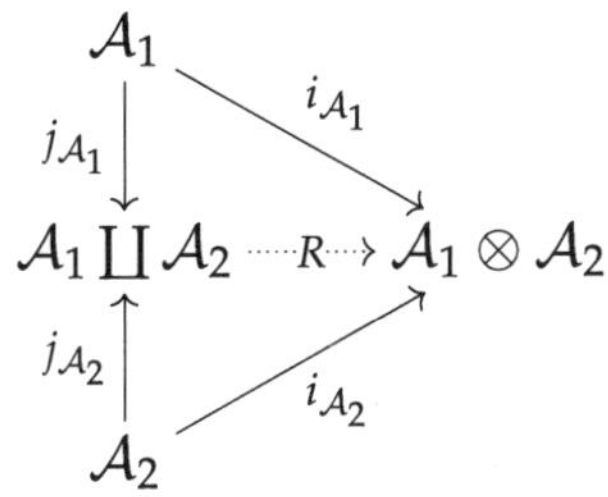

commute. The product

$$\big((\mathcal{A}_1, \varphi_1), (\mathcal{A}_2, \varphi_2)\big) \mapsto \big((\mathcal{A}_1 \coprod \mathcal{A}_2, \varphi_1 \tilde{\otimes} \varphi_2), j_{\mathcal{A}_1}, j_{\mathcal{A}_2}\big)$$

is called the *universal* version of the tensor product of functionals.

Example 1.8.11 (Free independence)
We will now introduce another product with inclusions for the category of algebraic probability spaces $\mathfrak{AlgProb}$. On the algebras we take simply the free product of algebras with identifications of units introduced in Example 1.8.2. This is the coproduct in the category of algebras; therefore, we also have natural inclusions. It only remains to define a unital linear functional on the free product of the algebras.

Voiculescu's[VDN92] *free product* $\varphi_1 * \varphi_2 : \mathcal{A}_1 \coprod \mathcal{A}_2 \to \mathbb{C}$ of two unital linear functionals $\varphi_1 : \mathcal{A}_1 \to \mathbb{C}$ and $\varphi_2 : \mathcal{A}_2 \to \mathbb{C}$ can be defined recursively by

$$\begin{aligned}(\varphi_1 * \varphi_2)(a_1 a_2 \cdots a_m) \\ = \sum_{I \subsetneq \{1,\ldots,m\}} (-1)^{m-\sharp I+1} (\varphi_1 * \varphi_2)\left(\overrightarrow{\prod_{k\in I}} a_k\right) \prod_{k \notin I} \varphi_{\epsilon_k}(a_k)\end{aligned} \tag{1.8.2}$$

for a typical element $a_1 a_2 \cdots a_m \in \mathcal{A}_1 \coprod \mathcal{A}_2$, with $a_k \in \mathcal{A}_{\epsilon_k}$, $\epsilon_1 \neq \epsilon_2 \neq \cdots \neq \epsilon_m$, that is, neighboring a's do not belong to the same algebra. Here $\sharp I$ denotes the number of elements of I and $\overrightarrow{\prod}_{k\in I} a_k$ means that the a's are to be multiplied in the same order in which they appear on the left-hand side. We use the convention $(\varphi_1 * \varphi_2)\left(\overrightarrow{\prod}_{k\in\emptyset} a_k\right) = 1$.

The formula in Equation (1.8.2) can be derived from Definition (1.7.2); it is the only way to define a functional $\Phi : \mathcal{A}_1 \coprod \mathcal{A}_2 \to \mathbb{C}$ such that the restrictions of Φ to the subalgebras $\mathcal{A}_i \cong j_{\mathcal{A}_i}(\mathcal{A}_i)$ agree with φ_i for $i = 1, 2$, and that these two subalgebras are free w.r.t. Φ.

Exercise 1.8.12 Derive Formula (1.8.2) from Definition 1.7.2 by expanding

$$0 = (\varphi_1 * \varphi_2)\left(\prod_{i=1}^{k} \left(a_i - \varphi_{\varepsilon_i}(a_i)\mathbf{1}\right)\right)$$

for $k \geq 1$, $\varepsilon \in \mathbb{A}_k$, and $a_1 \in \mathcal{A}_{\varepsilon_1}, \ldots, a_k \in \mathcal{A}_{\varepsilon_k}$.

If φ_1 and φ_2 are states, then their free product is a again a state, see also [VDN92, BNT05, NS06] and the references given there.

Example 1.8.13 (Boolean, monotone, and anti-monotone independence)

Ben Ghorbal and Schürmann[BG01, BGS99] and Muraki[Mur03] also considered the category of non-unital algebraic probability spaces, which we will denote by $\mathfrak{NuAlgProb}$, consisting of pairs $(\mathcal{A}, \varphi)$ of a not necessarily unital algebra $\mathcal{A}$ and a linear functional φ, in their classifications. On this category we can define three more products, namely the boolean product $\diamond$, the monotone product $\triangleright$ and the anti-monotone product $\triangleleft$, of states. They are defined by setting,

$$\varphi_1 \diamond \varphi_2(a_1 a_2 \cdots a_m) = \prod_{k=1}^{m} \varphi_{\epsilon_k}(a_k),$$

$$\varphi_1 \triangleright \varphi_2(a_1 a_2 \cdots a_m) = \varphi_1\left(\overrightarrow{\prod_{k:\epsilon_k=1}} a_k\right) \prod_{k:\epsilon_k=2} \varphi_2(a_k),$$

$$\varphi_1 \triangleleft \varphi_2(a_1 a_2 \cdots a_m) = \prod_{k:\epsilon_k=1} \varphi_1(a_k)\, \varphi_2\left(\overrightarrow{\prod_{k:\epsilon_k=2}} a_k\right),$$

for $\varphi_1 : \mathcal{A}_1 \to \mathbb{C}$ and $\varphi_2 : \mathcal{A}_2 \to \mathbb{C}$ and a typical element $a_1 a_2 \cdots a_m \in \mathcal{A}_1 \coprod \mathcal{A}_2$, $a_k \in \mathcal{A}_{\epsilon_k}$, $\epsilon_1 \neq \epsilon_2 \neq \cdots \neq \epsilon_m$, that is, neighboring a's don't belong to the same algebra. Note that we again have canonical inclusions $\mathcal{A}_i \to (\mathcal{A}_1 \coprod \mathcal{A}_2)$, $i = 1, 2$, since the free product without units is the coproduct in the category of not necessarily unital algebras.

The monotone and anti-monotone product are not commutative, but related by

$$\varphi_1 \triangleright \varphi_2 = (\varphi_2 \triangleleft \varphi_1) \circ \gamma_{\mathcal{A}_1, \mathcal{A}_2},$$

for all linear functionals $\varphi_1 : \mathcal{A}_1 \to \mathbb{C}$, $\varphi_2 : \mathcal{A}_2 \to \mathbb{C}$, where $\gamma_{\mathcal{A}_1, \mathcal{A}_2} : \mathcal{A}_1 \coprod \mathcal{A}_2 \to \mathcal{A}_2 \coprod \mathcal{A}_1$ is the commutativity constraint (for the commutativity constraint for the free product of unital algebras see Equation (1.8.1)). The boolean product is commutative, that is, it satisfies

$$\varphi_1 \diamond \varphi_2 = (\varphi_2 \diamond \varphi_1) \circ \gamma_{\mathcal{A}_1, \mathcal{A}_2},$$

for all linear functionals $\varphi_1 : \mathcal{A}_1 \to \mathbb{C}$, $\varphi_2 : \mathcal{A}_2 \to \mathbb{C}$.

Exercise 1.8.14 The boolean, the monotone, and the anti-monotone products can also be defined for unital algebras, if they are in the range of the unitization functor introduced in Exercise 1.8.5.

Let $\varphi_1 : \mathcal{A}_1 \to \mathbb{C}$ and $\varphi_2 : \mathcal{A}_2 \to \mathbb{C}$ be two unital functionals on algebras $\mathcal{A}_1$, $\mathcal{A}_2$, which can be decomposed as $\mathcal{A}_1 = \mathbb{C}\mathbf{1} \oplus \mathcal{A}_1^0$, $\mathcal{A}_2 = \mathbb{C}\mathbf{1} \oplus \mathcal{A}_2^0$. Then we define the boolean, monotone, or anti-monotone product of φ_1 and φ_2 as the unital extension of the boolean, monotone, or anti-monotone product of their restrictions $\varphi_1|_{\mathcal{A}_1^0}$ and $\varphi_2|_{\mathcal{A}_2^0}$.

This leads again to the formulae

$$\begin{aligned}
\varphi_1 \diamond \varphi_2(a_1 a_2 \cdots a_n) &= \prod_{i=1}^{n} \varphi_{\epsilon_i}(a_i), \\
\varphi_1 \triangleright \varphi_2(a_1 a_2 \cdots a_n) &= \varphi_1\left(\prod_{i:\,\epsilon_i=1} a_i\right) \prod_{i:\,\epsilon_i=2} \varphi_2(a_i), \\
\varphi_1 \triangleleft \varphi_2(a_1 a_2 \cdots a_n) &= \prod_{i:\,\epsilon_i=1} \varphi_1(a_i)\varphi_2\left(\prod_{i:\,\epsilon_i=2} a_i\right),
\end{aligned}$$

for $a_1 a_2 \cdots a_n \in \mathcal{A}_1 \coprod \mathcal{A}_2$, $a_i \in \mathcal{A}^0_{\epsilon_i}$, $\epsilon_1 \neq \epsilon_2 \neq \cdots \neq \epsilon_n$; however, note that now the factors a_i have to be chosen from $\mathcal{A}^0_{\epsilon_i}$. We use the convention that the empty product is equal to the unit element.

These products can be defined in the same way for $*$-algebraic probability spaces, where the algebras are unital $*$-algebras having such a decomposition $\mathcal{A} = \mathbb{C}\mathbf{1} \oplus \mathcal{A}_0$ and the functionals are states. To check that $\varphi_1 \diamond \varphi_2, \varphi_1 \triangleright \varphi_2, \varphi_1 \triangleleft \varphi_2$ are again states, if φ_1 and φ_2 are states, one can verify that the following constructions give their GNS representations. Let (π_1, H_1, ξ_1) and (π_2, H_2, ξ_2) denote the GNS representations of $(\mathcal{A}_1, \varphi_1)$ and $(\mathcal{A}_2, \varphi_2)$. The GNS representations of $(\mathcal{A}_1 \coprod \mathcal{A}_2, \varphi_1 \diamond \varphi_2)$, $(\mathcal{A}_1 \coprod \mathcal{A}_2, \varphi_1 \triangleright \varphi_2)$, and $(\mathcal{A}_1 \coprod \mathcal{A}_2, \varphi_1 \triangleleft \varphi_2)$ can all be defined on the Hilbert space $H = H_1 \otimes H_2$ with the state vector $\xi = \xi_1 \otimes \xi_2$. The representations are defined by $\pi(\mathbf{1}) = \mathrm{id}$ and

$$\begin{array}{llllll}
\pi|_{\mathcal{A}_1^0} = \pi_1 \otimes P_2, & \pi|_{\mathcal{A}_2^0} = P_1 \otimes \pi_2, & \text{for} & \varphi_1 \diamond \varphi_2, \\
\pi|_{\mathcal{A}_1^0} = \pi_1 \otimes P_2, & \pi|_{\mathcal{A}_2^0} = \mathrm{id}_{H_2} \otimes \pi_2, & \text{for} & \varphi_1 \triangleright \varphi_2, \\
\pi|_{\mathcal{A}_1^0} = \pi_1 \otimes \mathrm{id}_{H_2}, & \pi|_{\mathcal{A}_2^0} = P_1 \otimes \pi_2, & \text{for} & \varphi_1 \triangleleft \varphi_2,
\end{array}$$

where P_1, P_2 denote the orthogonal projections $P_1 : H_1 \to \mathbb{C}\xi_1$, $P_2 : H_2 \to \mathbb{C}\xi_2$. For the boolean case, $\xi = \xi_1 \otimes \xi_2 \in H_1 \otimes H_2$ is not cyclic for π, only the subspace $\mathbb{C}\xi \oplus H_1^0 \oplus H_2^0$ can be generated from ξ.

1.8.4 Classification of the universal independences

We will consider products of the form

$$((\mathcal{A}_1, \varphi_1), (\mathcal{A}_2, \varphi_2)) \mapsto ((\mathcal{A}_1 \coprod \mathcal{A}_2, \varphi_1 \cdot \varphi_2), i_{\mathcal{A}_1}, i_{\mathcal{A}_2})$$

on categories of algebraic probability spaces satisfying the following conditions.

- Associativity:

$$\big((\varphi_1 \cdot \varphi_2) \cdot \varphi_3\big) \circ \alpha_{\mathcal{A}_1,\mathcal{A}_2,\mathcal{A}_3} = \varphi_1 \cdot (\varphi_2 \cdot \varphi_3), \qquad \text{(Assoc)}$$

  for all $(\mathcal{A}_1, \varphi_1), (\mathcal{A}_2, \varphi_2), (\mathcal{A}_3, \varphi_3)$ in $\mathfrak{AlgProb}$, where $\alpha_{\mathcal{A}_1,\mathcal{A}_2,\mathcal{A}_3}$ denotes the canonical associativity morphism of the free product of (unital or non-unital) algebras, see Equation (1.8.1).
- Functoriality: we want the product to be a functor; therefore, the free product $j_1 \coprod j_2$ of two morphisms should again be a morphism. This is the case if and only if

$$(\varphi_1 \cdot \varphi_2) \circ (j_1 \coprod j_2) = (\varphi_1 \circ j_1) \cdot (\varphi_2 \circ j_2) \qquad \text{(Funct)}$$

  for all pairs of morphisms $j_1 : (\mathcal{B}_1, \psi_1) \to (\mathcal{A}_1, \varphi_1)$, $j_2 : (\mathcal{B}_2, \psi_2) \to (\mathcal{A}_2, \varphi_2)$.
- Inclusion: the inclusions $i_{\mathcal{A}_i} : \mathcal{A}_i \to (\mathcal{A}_1 \coprod \mathcal{A}_2)$, $i = 1, 2$, should be random variables (that is, morphisms) in the corresponding category. This is the case if and only if

$$(\varphi_1 \cdot \varphi_2) \circ i_{\mathcal{A}_1} = \varphi_1 \quad \text{and} \quad (\varphi_1 \cdot \varphi_2) \circ i_{\mathcal{A}_2} = \varphi_2 \qquad \text{(Incl)}$$

  for all pairs of objects $(\mathcal{A}_1, \varphi_1), (\mathcal{A}_2, \varphi_2)$.
- Normalization: in addition, we can also consider the condition

$$(\varphi_1 \cdot \varphi_2)(a_1 a_2) = \varphi_{\epsilon_1}(a_1)\varphi_{\epsilon_2}(a_2) \qquad \text{(Norm)}$$

  for all $(\epsilon_1, \epsilon_2) \in \big\{(1,2), (2,1)\big\}$, $a_1 \in \mathcal{A}_{\epsilon_1}$, $a_2 \in \mathcal{A}_{\epsilon_2}$.

Our Conditions (Assoc), (Incl), and (Funct) are exactly the axioms (P2), (P3), and (P4) in Ben Ghorbal and Schürmann[BGS99], or the axioms (U2), the first part of (U4), and (U3) in Muraki[Mur03].

The Conditions (Assoc), (Incl), and (Funct) can be reformulated; they express the fact that the product should be a tensor product equipped with natural inclusions of the objects into their product, *cf.* [Fra02, Fra06b, GL14, Ger15, Lac15]. We will call products satisfying (Assoc), (Incl), and (Funct) *tensor products with inclusions* or simply *universal products*.

Theorem 1.8.15 *(Muraki[Mur03], Ben Ghorbal, and Schürmann [BG01, BGS99]) There exist exactly two universal products on the category of algebraic probability spaces* $\mathfrak{AlgProb}$*, namely the universal* *version $\tilde{\otimes}$ of the tensor product defined in Example 1.8.10 and the free product $*$ of states defined in Example 1.8.11.*

In the non-unital case there exist more universal products.

Theorem 1.8.16 *(Muraki[Mur03]) There exist exactly five universal products satisfying (Norm) on the category of non-unital algebraic probability spaces $\mathfrak{NuAlgProb}$, namely the universal version $\tilde{\otimes}$ of the tensor product, the free product $*$, the boolean product $\diamond$, the monotone product $\triangleright$, and the anti-monotone product $\triangleleft$.*

The monotone and the anti-monotone product are not symmetric, that is, $(\mathcal{A}_1 \coprod \mathcal{A}_2, \varphi_1 \triangleright \varphi_2)$ and $(\mathcal{A}_2 \coprod \mathcal{A}_1, \varphi_2 \triangleright \varphi_1)$ are not isomorphic in general. Actually, the anti-monotone product is simply the mirror image of the monotone product,

$$(\mathcal{A}_1 \coprod \mathcal{A}_2, \varphi_1 \triangleright \varphi_2) \cong (\mathcal{A}_2 \coprod \mathcal{A}_1, \varphi_2 \triangleleft \varphi_1)$$

for all $(\mathcal{A}_1, \varphi_1), (\mathcal{A}_2, \varphi_2)$ in the category of non-unital algebraic probability spaces. The other three products are symmetric.

Condition (Norm) is not essential. If one drops it and adds symmetry, one finds in addition the degenerate product

$$(\varphi_1 \bullet_0 \varphi_2)(a_1 a_2 \cdots a_m) = \begin{cases} \varphi_{\epsilon_1}(a_1) & \text{if} \quad m = 1, \\ 0 & \text{if} \quad m > 1. \end{cases} \tag{1.8.3}$$

and families

$$\varphi_1 \bullet_q \varphi_2 = q\big((q^{-1}\varphi_1) \cdot (q^{-1}\varphi_2)\big), \tag{1.8.4}$$

parametrized by a complex number $q \in \mathbb{C}\backslash\{0\}$, for each of the three symmetric products, $\bullet \in \{\tilde{\otimes}, *, \diamond\}$.

If one adds the condition that products of states are again states, then one can also show that the constant has to be equal to one.

Exercise 1.8.17 Consider the category of non-unital $*$-algebraic probability spaces, whose objects are pairs $(\mathcal{A}, \varphi)$ consisting of a not necessarily unital $*$-algebra $\mathcal{A}$ and a state $\varphi : \mathcal{A} \to \mathbb{C}$. Here a state is a linear functional $\varphi : \mathcal{A} \to \mathbb{C}$ whose unital extension $\tilde{\varphi} : \tilde{\mathcal{A}} \cong \mathbb{C}\mathbf{1} \oplus \mathcal{A} \to \mathbb{C}$, $\lambda\mathbf{1} + a \mapsto \tilde{\varphi}(\lambda\mathbf{1} + a) = \lambda + \varphi(a)$, to the unitization of $\mathcal{A}$ is a state.

Assume we have products $\cdot : S(\mathcal{A}_1) \times S(\mathcal{A}_2) \to S(\mathcal{A}_1 \coprod \mathcal{A}_2)$ of linear functionals on non-unital algebras $\mathcal{A}_1, \mathcal{A}_2$ that satisfy

$$(\varphi_1 \cdot \varphi_2)(a_1 a_2) = c_1 \varphi_1(a_1)\varphi_2(a_2), \tag{1.8.5}$$
$$(\varphi_1 \cdot \varphi_2)(a_2 a_1) = c_2 \varphi_1(a_1)\varphi_2(a_2), \tag{1.8.6}$$

for all linear functionals $\varphi_1 : \mathcal{A}_1 \to \mathbb{C}$, $\varphi_2 : \mathcal{A}_2 \to \mathbb{C}$, and elements $a_1 \in \mathcal{A}_1$, $a_2 \in \mathcal{A}_2$ with 'universal scaling constants' $c_1, c_2 \in \mathbb{C}$, that is, constants that do not depend on the algebras, the functionals, or the algebra elements. That for every universal independence such constants have to exist is part of the proof of the classifications in [BG01, BGS99, Mur03].

Show that if the products of states are again states, then we have $c_1 = c_2 = 1$. Hint: Take for $\mathcal{A}_1$ and $\mathcal{A}_2$ the algebra of polynomials on $\mathbb{R}$ and for φ_1 and φ_2 evaluation in a point.

Even without symmetry one can show that there always exist constants c_1, c_2 as in Equations (1.8.5) and (1.8.6). If one imposes positivity, then it follows that $c_1 = c_2 = 1$, as in the Exercise above. In general, if these two constants are equal and not zero, one has the families

$$\varphi_1 \bullet_q \varphi_2 = q\big((q^{-1}\varphi_1) \cdot (q^{-1}\varphi_2)\big),$$

parametrized by a complex number $q \in \mathbb{C}\backslash\{0\}$, for each of the five products, $\bullet \in \{\tilde{\otimes}, *, \diamond, \triangleright, \triangleleft\}$, similar to the situation in the symmetric case described in Equation (1.8.4). If the constants are not equal, a new family of products, called (r,s)-products, arise, see [GL14, Ger15, Lac15]. It is conjectured that the degenerate product in Equation (1.8.3) is the only possibility in the case $c_1 = c_2 = 0$.

The proof of the classification of universal independences can be split into three steps.

Using the 'universality' or functoriality of the product, one can show that there exist some 'universal combinatorial constants' - not depending on the algebras - and a formula for evaluating

$$(\varphi_1 \cdot \varphi_2)(a_1 a_2 \cdots a_m)$$

for $a_1 a_2 \cdots a_m \in \mathcal{A}_1 \coprod \mathcal{A}_2$, with $a_k \in \mathcal{A}_{\epsilon_k}$, $\epsilon_1 \neq \epsilon_2 \neq \cdots \neq \epsilon_m$, as a linear combination of products $\varphi_1(M_1)$, $\varphi_2(M_2)$, where M_1, M_2 are 'sub-monomials' of $a_1 a_2 \cdots a_m$. Then, in the second step, it is shown by associativity that only products with *ordered* monomials M_1, M_2 contribute. This is the content of [BGS02, Theorem 5] in the commutative case and of [Mur03, Theorem 2.1] in the general case.

The third step, which was actually completed first in both cases, see [Spe97] and [Mur02], is to find the conditions that the universal constants have to satisfy, if the resulting product is associative. It turns out that the universal coefficients for $m > 5$ are already uniquely determined by the coefficients for $1 \leq m \leq 5$. Detailed analysis of the non linear equations obtained for the coefficients of order up to five then leads to the classifications stated above.

Muraki [Mur13] has recently shown that a simpler proof is possible if one imposes positivity.

1.9 Lévy Processes on Dual Groups

We now want to study quantum stochastic processes whose increments are free or independent in the sense of boolean, monotone, or anti-monotone independence. The approach based on bialgebras that we followed in the Section 1.5 and 1.6 works for the tensor product and fails in the other cases because the corresponding products are not defined on the tensor product, but on the free product of the algebra. The algebraic structures that have to replace bialgebras were introduced into quantum probability by Voiculescu [Voi87, Voi90], who named them dual groups. Closely related structures have also been studied in other contexts under the name H-algebra [Zha91] or co-group [Ber65, BH96, Fre98].

In this section we will introduce these algebras and present the theory of their Lévy processes. Lévy processes on dual groups with boolean, monotonically, or anti-monotonically independent increments can be reduced to Lévy processes on involutive bialgebra, *cf.* [Fra03a]. We do not know if this is also possible for Lévy processes on dual groups with free increments.

In the literature additive free Lévy processes have been studied most intensively, see, for example, [GSS92, Bia98, Ans02, Ans03] [BNT02b, BNT02a, Vos13, Ulr14].

1.9.1 Dual groups

Denote by $\mathfrak{ComAlg}$ the category of commutative unital algebras and let $\mathcal{B} \in \mathrm{Ob}\,\mathfrak{ComAlg}$ be a commutative bialgebra. Then the mapping

$$\mathrm{Mor}_{\mathfrak{ComAlg}}(\mathcal{B}, \cdot) : \mathrm{Ob}\,\mathfrak{ComAlg} \ni \mathcal{A} \mapsto \mathrm{Mor}_{\mathfrak{ComAlg}}(\mathcal{B}, \mathcal{A})$$

can be understood as a functor from $\mathfrak{ComAlg}$ to the category of unital semigroups. The multiplication in $\mathrm{Mor}_{\mathfrak{Alg}}(\mathcal{B},\mathcal{A})$ is given by the convolution, that is,

$$f \star g = m_{\mathcal{A}} \circ (f \otimes g) \circ \Delta_{\mathcal{B}}$$

and the unit element is $\varepsilon_{\mathcal{B}}\mathbf{1}_{\mathcal{A}} : \mathcal{B} \ni b \mapsto \varepsilon_{\mathcal{B}}(b)\mathbf{1}_{\mathcal{A}} \in \mathcal{A}$. A unit-preserving algebra homomorphism $h : \mathcal{A}_1 \to \mathcal{A}_2$ gets mapped to the unit-preserving semigroup homomorphism

$$\mathrm{Mor}_{\mathfrak{ComAlg}}(\mathcal{B},\mathcal{A}_1) \ni f \to h \circ f \in \mathrm{Mor}_{\mathfrak{ComAlg}}(\mathcal{B},\mathcal{A}_2),$$

since

$$\begin{aligned} h \circ (\varepsilon_{\mathcal{B}}\mathbf{1}_{\mathcal{A}_1}) &= \varepsilon_{\mathcal{B}}\mathbf{1}_{\mathcal{A}_2}, \\ h \circ (f \star g) &= (h \circ f) \star (h \circ g), \end{aligned}$$

for $\mathcal{A}_1, \mathcal{A}_2 \in \mathrm{Ob}\,\mathfrak{ComAlg}$, $h \in \mathrm{Mor}_{\mathfrak{ComAlg}}(\mathcal{A}_1,\mathcal{A}_2)$, and algebra homomorphisms $f,g \in \mathrm{Mor}_{\mathfrak{ComAlg}}(\mathcal{B},\mathcal{A}_1)$.

If $\mathcal{B}$ is a commutative Hopf algebra with antipode S, then the functor $\mathrm{Mor}_{\mathfrak{ComAlg}}(\mathcal{B},\cdot)$ takes values in the category of groups, since then $\mathrm{Mor}_{\mathfrak{ComAlg}}(\mathcal{B},\mathcal{A})$ is a group with respect to the convolution product. The inverse of a homomorphism $f : \mathcal{B} \to \mathcal{A}$ with respect to the convolution product is given by $f \circ S$.

The calculation

$$\begin{aligned} (f \star g)(ab) &= m_{\mathcal{A}} \circ (f \otimes g) \circ \Delta_{\mathcal{B}}(ab) \\ &= f(a_{(1)}b_{(1)})g(a_{(2)}b_{(2)}) = f(a_{(1)})f(b_{(1)})g(a_{(2)})g(b_{(2)}) \\ &= f(a_{(1)})g(a_{(2)})f(b_{(1)})g(b_{(2)}) = (f \star g)(a)(f \star g)(b) \end{aligned}$$

shows that the convolution product $f \star g$ of two homomorphisms $f,g : \mathcal{B} \to \mathcal{A}$ is again a homomorphism. It also gives an indication why noncommutative bialgebras or Hopf algebras do not give rise to a similar functor on the category of noncommutative algebras, since we had to commute $f(b_{(1)})$ with $g(a_{(2)})$.

Zhang [Zha91], Bergman, and Hausknecht [BH96] showed that if one replaces the tensor product in the definition of bialgebras and Hopf algebras by the free product, then one arrives at a class of algebras that do give rise to a functor from the category of noncommutative algebras to the category of semigroups or groups.

A *dual group* [Voi87, Voi90] (called *H-algebra* or *cogroup* in the category of unital associative $*$-algebras in [Zha91] and [BH96], resp.) is a unital $*$-algebra $\mathcal{B}$ equipped with three unital $*$-algebra homomorphisms $\Delta : \mathcal{B} \to \mathcal{B} \coprod \mathcal{B}$, $S : \mathcal{B} \to \mathcal{B}$, and $\varepsilon : \mathcal{B} \to \mathbb{C}$ (also called *comultiplication or coproduct, antipode,* and *counit*) such that

$$\left(\Delta \coprod \mathrm{id}\right) \circ \Delta \;=\; \left(\mathrm{id} \coprod \Delta\right) \circ \Delta, \tag{1.9.1}$$

$$\left(\varepsilon \coprod \mathrm{id}\right) \circ \Delta = \;\mathrm{id}\; = \left(\mathrm{id} \coprod \varepsilon\right) \circ \Delta, \tag{1.9.2}$$

$$m_{\mathcal{B}} \circ \left(S \coprod \mathrm{id}\right) \circ \Delta = \;\mathrm{id}\; = m_{\mathcal{B}} \circ \left(\mathrm{id} \coprod S\right) \circ \Delta, \tag{1.9.3}$$

where $m_{\mathcal{B}} : \mathcal{B} \coprod \mathcal{B} \to \mathcal{B}$, $m_{\mathcal{B}}(a_1 \otimes a_2 \otimes \cdots \otimes a_n) = a_1 \cdot a_2 \cdot \cdots \cdot a_n$, is the multiplication of $\mathcal{B}$. Besides the formal similarity, there are many relations between dual groups on the one side and Hopf algebras and bialgebras on the other side, *cf.* [Zha91]. For example, let $\mathcal{B}$ be a dual group with comultiplication Δ, and let $R : \mathcal{B} \coprod \mathcal{B} \to \mathcal{B} \otimes \mathcal{B}$ be the unique unital $*$-algebra homomorphism with

$$R_{\mathcal{B},\mathcal{B}} \circ i_1(b) = b \otimes \mathbf{1}, \qquad R_{\mathcal{B},\mathcal{B}} \circ i_2(b) = \mathbf{1} \otimes b,$$

for all $b \in \mathcal{B}$. Here $i_1, i_2 : \mathcal{B} \to \mathcal{B} \coprod \mathcal{B}$ denote the canonical inclusions of $\mathcal{B}$ into the first and the second factors of the free product $\mathcal{B} \coprod \mathcal{B}$, respectively. Then $\mathcal{B}$ is a bialgebra with the comultiplication $\overline{\Delta} = R_{\mathcal{B},\mathcal{B}} \circ \Delta$, see [Zha91, Theorem 4.2], but in general it is not a Hopf algebra.

We will not really work with dual groups, but with the following weaker notion. A *dual semigroup* is a unital $*$-algebra $\mathcal{B}$ equipped with two unital $*$-algebra homomorphisms $\Delta : \mathcal{B} \to \mathcal{B} \coprod \mathcal{B}$ and $\varepsilon : \mathcal{B} \to \mathbb{C}$ such that Equations (1.9.1) and (1.9.2) are satisfied. The antipode is not used in the proof of [Zha91, Theorem 4.2], and therefore we also get an involutive bialgebra $(\mathcal{B}, \overline{\Delta}, \varepsilon)$ for every dual semigroup $(\mathcal{B}, \Delta, \varepsilon)$.

Note that we can always write a dual semigroup $\mathcal{B}$ as a direct sum $\mathcal{B} = \mathbb{C}\mathbf{1} \oplus \mathcal{B}^0$, where $\mathcal{B}^0 = \ker \varepsilon$ is even a $*$-ideal. Therefore, it is in the range of the unitization functor and the boolean, monotone, and anti-monotone product can be defined for unital linear functionals on $\mathcal{B}$, *cf.* Exercise 1.8.14.

The comultiplication of a dual semigroup can also be used to define a convolution product. The *convolution* $j_1 \star j_2$ of two unital $*$-algebra homomorphisms $j_1, j_2 : \mathcal{B} \to \mathcal{A}$ is defined as

$$j_1 \star j_2 = m_{\mathcal{A}} \circ \left(j_1 \coprod j_2\right) \circ \Delta.$$

As the composition of the three unital $*$-algebra homomorphisms $\Delta : \mathcal{B} \to \mathcal{B} \coprod \mathcal{B}$, $j_1 \coprod j_2 : \mathcal{B} \coprod \mathcal{B} \to \mathcal{A} \coprod \mathcal{A}$, and $m_{\mathcal{A}} : \mathcal{A} \coprod \mathcal{A} \to \mathcal{A}$, this is obviously again a unital $*$-algebra homomorphism. The convolution $j_1 \star j_2$ should not be thought of as an analog of the convolution of probability distributions, but rather as an analog of the composition of random variables by pointwise multiplication. It is this operation that we will use to give a sense to the notion of increments of a stochastic process.

Note that this convolution can not be defined for arbitrary linear maps on $\mathcal{B}$ with values in some algebra, as for bialgebras, but only for unital $*$-algebra homomorphisms.

1.9.2 Definition of Lévy processes on dual groups

Definition 1.9.1 Let $j_1 : \mathcal{B}_1 \to (\mathcal{A}, \Phi), \ldots, j_n : \mathcal{B}_n \to (\mathcal{A}, \Phi)$ be quantum random variables over the same quantum probability space $(\mathcal{A}, \Phi)$ and denote their marginal distributions by $\varphi_i = \Phi \circ j_i$, $i = 1, \ldots, n$. The quantum random variables $(j_1, \ldots, j_n)$ are called tensor-independent (respectively, boolean-independent, monotonically independent, anti-monotonically independent, or free), if the state $\Phi \circ m_{\mathcal{A}} \circ (j_1 \coprod \cdots \coprod j_n)$ on the free product $\coprod_{i=1}^{n} \mathcal{B}_i$ is equal to the tensor product (boolean, monotone, anti-monotone, or free product, respectively) of $\varphi_1, \ldots, \varphi_n$.

Exercise 1.9.2 Derive Definition 1.9.1 from Definition 1.8.6 (for $n = 2$).

Note that tensor, boolean, and free independence do not depend on the order, but monotone and anti-monotone independence do. An n-tuple $(j_1, \ldots, j_n)$ of quantum random variables is monotonically independent, if and only if $(j_n, \ldots, j_1)$ is anti-monotonically independent.

We are now ready to define tensor, boolean, monotone, anti-monotone, and free Lévy processes on dual semigroups.

Definition 1.9.3 [Sch95b] Let $(\mathcal{B}, \Delta, \varepsilon)$ be a dual semigroup. A quantum stochastic process $\{j_{st}\}_{0 \leq s \leq t \leq \infty}$ on $\mathcal{B}$ over some quantum probability space $(\mathcal{A}, \Phi)$ is called a *tensor (resp. boolean, monotone, anti-monotone, or free) Lévy process on the dual semigroup* $\mathcal{B}$, if the following four conditions are satisfied.

(i) Increment property: We have

$$\begin{aligned} j_{rs} \star j_{st} &= j_{rt} \quad \text{for all } 0 \le r \le s \le t \le \infty, \\ j_{tt} &= \varepsilon \mathbf{1}_{\mathcal{A}} \quad \text{for all } 0 \le t \le \infty. \end{aligned}$$

(ii) Independence of increments: The family $\{j_{st}\}_{0\le s\le t\le\infty}$ is tensor-independent (resp. boolean, monotonically, anti-monotonically independent, or free) w.r.t. Φ, that is, the n-tuple $(j_{s_1t_2},\ldots,j_{s_nt_n})$ is tensor-independent (resp. boolean, monotonically, anti-monotonically independent, or free) for all $n \in \mathbb{N}$ and all $0 \le s_1 \le t_1 \le s_2 \le \cdots \le t_n \le \infty$.

(iii) Stationarity of increments: The distribution $\varphi_{st} = \Phi \circ j_{st}$ of j_{st} depends only on the difference $t - s$.

(iv) Weak continuity: The quantum random variables j_{st} converge to j_{ss} in distribution for $t \searrow s$.

Remark 1.9.4 The independence property depends on the products and therefore for boolean, monotone, and anti-monotone Lévy processes on the choice of a decomposition $\mathcal{B} = \mathbb{C}\mathbf{1} \oplus \mathcal{B}^0$. In order to show that the convolutions defined by $(\varphi_1 \diamond \varphi_2) \circ \Delta$, $(\varphi_1 \triangleright \varphi_2) \circ \Delta$, and $(\varphi_1 \triangleleft \varphi_2) \circ \Delta$ are associative and that the counit ε acts as unit element w.r.t. these convolutions, one has to use the universal property [BGS99, Condition (P4)], which in our setting is only satisfied for morphisms that respect the decomposition. Therefore, we are forced to choose the decomposition given by $\mathcal{B}^0 = \ker \varepsilon$.

The marginal distributions $\varphi_{t-s} := \varphi_{st} = \Phi \circ j_{st}$ form again a convolution semigroup $\{\varphi_t\}_{t\in\mathbb{R}_+}$, with respect to the tensor (boolean, monotone, anti-monotone, or free, respectively) convolution defined by $(\varphi_1 \tilde{\otimes} \varphi_2) \circ \Delta$ ($(\varphi_1 \diamond \varphi_2) \circ \Delta$, $(\varphi_1 \triangleright \varphi_2) \circ \Delta$, $(\varphi_1 \triangleleft \varphi_2) \circ \Delta$, or $(\varphi_1 * \varphi_2) \circ \Delta$, respectively). It has been shown that the generating functional $\psi : \mathcal{B} \to \mathbb{C}$,

$$\psi(b) = \lim_{t\searrow 0} \frac{1}{t}\left(\varphi_t(b) - \varepsilon(b)\right)$$

is well defined for all $b \in \mathcal{B}$ and uniquely characterizes the semigroup $\{\varphi_t\}_{t\in\mathbb{R}_+}$, *cf.* [Sch95b, BGS99, Fra01].

Schürmann and Voss [SV14] have given a new proof of the Schoenberg correspondence, using ideas from [SSV10], that includes also the free convolution. This shows that Lévy processes on dual semigroups are in one-to-one correspondence with generating functionals, that is, hermitian, conditionally positive linear functionals that vanish on the identity.

Theorem 1.9.5 Schoenberg correspondence: *Let $\{\varphi_t\}_{t\geq 0}$ be a convolution semigroup of unital functionals with respect to the tensor, boolean, monotone, or anti-monotone convolution on a dual semigroup $(\mathcal{B},\Delta,\varepsilon)$ and let $\psi:\mathcal{B}\to\mathbb{C}$ be defined by*

$$\psi(b)=\lim_{t\searrow 0}\frac{1}{t}\big(\varphi_t(b)-\varepsilon(b)\big)$$

for $b\in\mathcal{B}$. Then the following statements are equivalent.

(i) φ_t is positive for all $t\geq 0$.

(ii) ψ is hermitian and conditionally positive.

Conversely, φ_t can be recovered from the generating functional ψ as a kind of convolution exponential (defined by extending ψ to the symmetric tensor algebra over $\mathcal{B}$). For details see [Sch95b, Theorem 3.3].

It is well known that freeness arises in the study of the behavior of random matrices for large dimension, see, e.g., [Voi91, Bia97] and the references therein. Ulrich has shown recently that free Lévy processes can also be obtained in this way. He has shown that the limit in distribution of the $d\times d$ blocks of Brownian motion in the unitary group $U(nd)$ as n tends to infinity is a free Lévy processes on the dual group analog of the unitary group, *cf.* [Ulr14].

1.9.3 Time reversal

Denote by F the *flip* map $F:\mathcal{B}\coprod\mathcal{B}\to\mathcal{B}\coprod\mathcal{B}, F=m_{\mathcal{B}\coprod\mathcal{B}}\circ(i_2\coprod i_1)$, where $i_1,i_2:\mathcal{B}\to\mathcal{B}\coprod\mathcal{B}$ are the inclusions of $\mathcal{B}$ into the first and the second factors of the free product $\mathcal{B}\coprod\mathcal{B}$. The flip map F acts on $i_1(a_1)i_2(b_1)\cdots i_2(b_n)\in\mathcal{B}\coprod\mathcal{B}$ with $a_1,\ldots,a_n,b_1,\ldots,b_n\in\mathcal{B}$ as

$$F\big(i_1(a_1)i_2(b_1)\cdots i_2(b_n)\big)=i_2(a_1)i_1(b_1)\cdots i_1(b_n).$$

If $j_1:\mathcal{B}\to\mathcal{A}_1$ and $j_2:\mathcal{B}\to\mathcal{A}_2$ are two unital $*$-algebra homomorphisms, then we have $(j_2\coprod j_1)\circ F=\gamma_{\mathcal{A}_1,\mathcal{A}_2}\circ(j_1\coprod j_2)$.

Like for bialgebras, the opposite comultiplication $\Delta^{\rm op} = F \circ \Delta$ of a dual semigroup $(\mathcal{B}, \Delta, \varepsilon)$ defines a new dual semigroup $(\mathcal{B}, \Delta^{\rm op}, \varepsilon)$.

In the following we restrict the index set to a finite interval $[0, T]$, with some fixed $T > 0$. Then we can define time reversal as the map $[0, T] \ni t \mapsto T - t \in [0, T]$. It is straight forward to adapt the definition of Lévy processes to finite time intervals.

Proposition 1.9.6 *Let $\{j_{st} : \mathcal{B} \to (\mathcal{A}, \Phi)\}_{0 \leq s \leq t \leq T}$ be a quantum stochastic process on a dual semigroup $(\mathcal{B}, \Delta, \varepsilon)$ and define its time-reversed process $\{j^{\rm op}_{st}\}_{0 \leq s \leq t \leq T}$ by*

$$j^{\rm op}_{st} = j_{T-t, T-s}$$

for $0 \leq s \leq t \leq \infty$.

(i) The process $\{j_{st}\}_{0 \leq s \leq t \leq T}$ is a tensor (boolean, free, respectively) Lévy process on the dual semigroup $(\mathcal{B}, \Delta, \varepsilon)$ if and only if the time-reversed process $\{j^{\rm op}_{st}\}_{0 \leq s \leq t \leq T}$ is a tensor (boolean, free, respectively) Lévy process on the dual semigroup $(\mathcal{B}, \Delta^{\rm op}, \varepsilon)$.

(ii) The process $\{j_{st}\}_{0 \leq s \leq t \leq T}$ is a monotone (anti-monotone, resp.) Lévy process on the dual semigroup $(\mathcal{B}, \Delta, \varepsilon)$ if and only if the time-reversed process $\{j^{\rm op}_{st}\}_{0 \leq s \leq t \leq T}$ is an anti-monotone (monotone, resp.) Lévy process on the dual semigroup $(\mathcal{B}, \Delta^{\rm op}, \varepsilon)$.

Proof The equivalence of the stationarity and continuity property for the quantum stochastic processes $\{j_{st}\}_{0 \leq s \leq t \leq T}$ and $\{j^{\rm op}_{st}\}_{0 \leq s \leq t \leq T}$ is clear.

The increment property for $\{j_{st}\}_{0 \leq s \leq t \leq T}$ with respect to Δ is equivalent to the increment property of $\{j^{\rm op}_{st}\}_{0 \leq s \leq t \leq T}$ with respect to $\Delta^{\rm op}$, since

$$\begin{aligned} m_{\mathcal{A}} \circ \left(j^{\rm op}_{st} \coprod j^{\rm op}_{tu} \right) \circ \Delta^{\rm op} &= m_{\mathcal{A}} \circ \left(j_{T-t,T-s} \coprod j_{T-u,T-t} \right) \circ F \circ \Delta \\ &= m_{\mathcal{A}} \circ \gamma_{\mathcal{A},\mathcal{A}} \circ \left(j_{T-u,T-t} \coprod j_{T-t,T-s} \right) \circ \Delta \\ &= m_{\mathcal{A}} \circ \left(j_{T-u,T-t} \coprod j_{T-t,T-s} \right) \circ \Delta \end{aligned}$$

for all $0 \leq s \leq t \leq u \leq T$.

If the process $\{j_{st}\}_{0 \leq s \leq t \leq T}$ has monotonically independent increments, that is, if the n-tuples $(j_{s_1 t_2}, \ldots, j_{s_n t_n})$ are monotonically independent for all $n \in \mathbb{N}$ and all $0 \leq s_1 \leq t_1 \leq s_2 \leq \cdots \leq t_n$, then the n-tuples $(j_{s_n t_n}, \ldots, j_{s_1 t_1}) = (j^{\rm op}_{T-t_n, T-s_n}, \ldots, j^{\rm op}_{T-t_1, T-s_1})$ are anti-monotonically independent and therefore

$\{j_{st}^{\rm op}\}_{0\leq s\leq t\leq T}$ has anti-monotonically independent increments, and vice versa.

Since tensor and boolean independence and freeness do not depend on the order, $\{j_{st}\}_{0\leq s\leq t\leq T}$ has tensor (boolean, free, respectively) independent increments, if and only if $\{j_{st}^{\rm op}\}_{0\leq s\leq t\leq T}$ has tensor (boolean, free, respectively) independent increments. □

1.10 Open Problems

We close this lecture with a list of interesting research topics and open questions.

(i) We defined Gaussian generating functionals in Proposition 1.5.13. Let L be a generating functional. We say that a Gaussian generating functional L_G is a Gaussian component of L, if $L - L_G$ is conditionally positive, and that L_G is maximal if $L_G - L'$ is conditionally positive for all Gaussian components L' of L. If it exists, then the maximal Gaussian component is unique up to a hermitian left invariant derivation, that is, a drift, see [Sch90a, Proposition 3.8]. Schürmann has shown that generating functionals on commutative involutive bialgebras or on the so-called 'noncommutative coefficient algebra of the unitary group' admit maximal Gaussian components, see [Sch90a] [Sch93, Section 5], but in general this question is still open.[1]

(ii) In Equation (1.6.11) we showed that any translation invariant Markov semigroup on a compact quantum group $\mathbb{G}$ is equal to the Markov semigroup of a Lévy process on its $*$-Hopf algebra, up to elements from N_h, the null space of the Haar state. Is it possible to remove N_h from Equation (1.6.11)?

(iii) There are other notions of quantum groups, whose structure can not be captured in involutive bialgebras as in Woronowicz' theory of compact quantum groups, and therefore Schürmann's theory of Lévy processes can not be applied directly. It would be interesting to develop a theory of Lévy processes for these quantum groups, too. There are in

[1]After these notes where written, there was some progress on this problem. Franz, Gerhold, and Thom showed that there exist generating functionals on the group algebra of fundamental groups of oriented surfaces of genus $k \geq 2$ that do not admit such a 'Lévy-Khintchine' decomposition into a maximal Gaussian part and a 'rest', *cf.* [FGT15].

particular Van Daele's ($*$-)algebraic quantum groups [VD98, VD03] and Kusterman and Vaes' locally compact quantum groups [KV99, KV00].

The case of Poisson-type Lévy processes on locally compact quantum groups has been worked out in [LS11], and some further initial work in this direction has been done in [LS12].

(iv) A construction that reduces the monotone, anti-monotone, and the boolean product of quantum probability spaces to the tensor product was introduced in [Fra03b]. This construction allows to construct monotone, anti-monotone, and boolean Lévy processes on dual groups from Lévy processes on $*$-bialgebras. A construction of free products from tensor products was given by Lenczewski [Len98]; however, it is not clear how to apply it to the theory of free Lévy processes. Does there exist a 'reduction' from freeness to tensor independence, or vice versa, similar to the construction in [Fra03b]?

(v) One question remains open in the classification of universal products of algebraic probability spaces: Is the degenerate universal product defined in Equation (1.8.3) the only universal product in the case where both constants c_1 and c_2 in Equations (1.8.5) and (1.8.6) are equal to zero? See [Lac15, Page 7].

(vi) It would be interesting to find new examples of dual groups and to study their properties. Voiculescu has defined dual group analogs of the classical series of simple Lie groups, *cf.* [Voi87], but only the dual group analog of the unitary groups has been studied in the literature, see [GvW89, McC92]. In quantum probability only the examples of the tensor algebra, the free group algebra, and the dual group analog of the unitary group have been studied, see, e.g, [Sch95b, BG01, Vos13].

(vii) Ulrich [Ulr14] has shown that the limit in distribution of the $d \times d$ blocks of Brownian motion in the unitary group $U(nd)$ as n tends to infinity is a free Lévy process on the dual group analog of the unitary group, *cf.* [Ulr14]. In [CU15] this result was generalized to a larger class of Lévy processes. It would be interesting to study if such a result also holds for orthogonal and symplectic groups and if it can be applied to get new insights into the behavior of large random matrices.

References

[AFS02] L. Accardi, U. Franz, and M. Skeide. Renormalized squares of white noise and other non-Gaussian noises as Lévy processes on real Lie algebras. *Comm. Math. Phys.*, 228(1): 123–150, 2002.

[ALS07] L. Accardi, R. Lenczewski, and R. Sałapata. Decompositions of the free product of graphs. *Infin. Dimens. Anal. Quantum Probab. Relat. Top.*, 10(3): 303–334, 2007.

[Ans02] M. Anshelevich. Itô formula for free stochastic integrals. *J. Funct. Anal.*, 188(1): 292–315, 2002.

[Ans03] M. Anshelevich. Free martingale polynomials. *J. Funct. Anal.*, 201(1): 228–261, 2003.

[App04] D. Applebaum. *Lévy Processes and Stochastic Calculus.* Cambridge University Press, 2004.

[App05] D. Applebaum. Lectures on classical Lévy process in Euclidean spaces and groups. In [QIIP-I], pp. 1–98, 2005.

[App10] D. Applebaum, Infinitely Divisible Central Probability Measures on Compact Lie Groups - Regularity, Semigroups and Transition Kernels. *Ann. Probab.* 39 (2011), no. 6, 2474–2496.

[App14] D. Applebaum. *Probability on Compact Lie Groups.* Stochastic Modelling and Probability Theory Vol. 70, Springer-Verlag, Berlin, 2014.

[Ara09] H. Araki. *Mathematical theory of quantum fields.* Oxford University Press, reprint of the 1999 hardback edition, 2009.

[ASW88] L. Accardi, M. Schürmann, and W.v. Waldenfels. Quantum independent increment processes on superalgebras. *Math. Z.*, 198: 451–477, 1988.

[Ban96] T. Banica. Théorie des représentations du groupe quantique compact libre $O(n)$. *C. R. Acad. Sci. Paris Sér. I Math.*, 322: 241–244, 1996.

[Bel98] V.P. Belavkin. On quantum Itô algebras. *Math. Phys. Lett.*, 7: 1–16, 1998.

[Bel08] S.T. Belinschi. The Lebesgue decomposition of the free additive convolution of two probability distributions. *Probab. Theory Related Fields* 142(1-2): 125–150, 2008.

[BB04] S.T. Belinschi and H. Bercovici. Atoms and regularity for measures in a partially defined free convolution semigroup. *Math. Z.* 248(4): 665–674, 2004.

[BB05] S.T. Belinschi and H. Bercovici. Partially defined semigroups relative to multiplicative free convolution. *Int. Math. Res. Not.*, 2005(2): 65–101, 2005.

[Ber65] I. Berstein. On co-groups in the category of graded algebras. *Trans. Amer. Math. Soc.* 115 (1965), 257–269.

[Ber05a] H. Bercovici. Multiplicative monotonic convolution. *Illinois J. Math.*, 49(3): 929–951 (electronic), 2005.

[Ber05b] H. Bercovici. A remark on monotonic convolution. *Infin. Dimens. Anal. Quantum Probab. Relat. Top.*, 8(1): 117–120, 2005.

[Ber06] H. Bercovici. On Boolean convolutions. In *Operator theory 20*, volume 6 of *Theta Ser. Adv. Math.*, pages 7–13. Theta, Bucharest, 2006.

[Ber98] J. Bertoin. *Lévy Processes.* Cambridge University Press, 1998.

[BG01] A. Ben Ghorbal. *Fondements algébrique des probabilités quantiques et calcul stochastique sur l'espace de Fock booléen.* PhD thesis, Université Henri Poincaré-Nancy 1, 2001.

[BGDS01] A. Ben Ghorbal and M. Schürmann. Quantum stochastic calculus on Boolean Fock space. *Infin. Dimens. Anal. Quantum Probab. Relat. Top.*, Vol. 7, No. 4, pp. 631–650, 2004.

[BGS99] A. Ben Ghorbal and M. Schürmann. On the algebraic foundations of a noncommutative probability theory. Prépublication 99/17, Institut E. Cartan, Nancy, 1999.

[BGS02] A. Ben Ghorbal and M. Schürmann. Noncommutative notions of stochastic independence. *Math. Proc. Cambridge Philos. Soc.*, 133(3): 531–561, 2002.

[BH96] G.M. Bergman and A.O. Hausknecht. *Co-groups and co-rings in categories of associative rings*, volume 45 of *Mathematical Surveys and Monographs*. American Mathematical Society, Providence, RI, 1996.

[Bha01] B.V.R. Bhat. Cocycles of CCR flows. *Mem. Amer. Math. Soc.*, 149(709), 2001.

[Bha05] B.V.R. Bhat. Dilations, cocycles, and product systems. In [QIIP-I], pp. 273–291, 2005.

[Bia97] P. Biane. Free Brownian motion, free stochastic calculus and random matrices. In:Free probability theory (Waterloo, ON, 1995), pp. 1–19, *Fields Inst. Commun.*, 12, Amer. Math. Soc., Providence, RI, 1997.

[Bia98] P. Biane. Processes with free increments. *Math. Z.*, 227(1): 143–174, 1998.

[BMT01] E. Bédos, G.J. Murphy, and L. Tuset. Co-amenability of compact quantum groups. *J. Geom. Phys.*, 40(2): 130–153, 2001.

[BMT03] E. Bédos, G.J. Murphy, and L. Tuset. Amenability and co-amenability of algebraic quantum groups. II. *J. Funct. Anal.*, 201(2): 303–340, 2003.

[BNT02a] O.E. Barndorff-Nielsen and S. Thorbjørnsen. Lévy laws in free probability. *Proc. Natl. Acad. Sci. USA*, 99(26): 16568–16575 (electronic), 2002.

[BNT02b] O.E. Barndorff-Nielsen and S. Thorbjørnsen. Self-decomposability and Lévy processes in free probability. *Bernoulli*, 8(3): 323–366, 2002.

[BNT05] O.E. Barndorff-Nielsen and S. Thorbjørnsen. On the roles of classical and free Lévy processes in theory and applications. In [QIIP-II], 2005.

[BR97] O. Bratteli and D.W. Robinson. *Operator algebras and quantum statistical mechanics. 2: Equilibrium states. Models in quantum statistical mechanics. 2nd ed.* Texts and Monographs in Physics. Berlin: Springer., 1997.

[Bra12] M. Brannan. Approximation properties for free orthogonal and free unitary quantum groups. *J. Reine Angew. Math.* 672 (2012), 223–251.

[BV92] H. Bercovici and D. Voiculescu. Lévy-Hinčin type theorems for multiplicative and additive free convolution. *Pacific J. Math.*, 153(2): 217–248, 1992.

[BV93] H. Bercovici and D. Voiculescu. Free convolution of measures with unbounded support. *Indiana Univ. Math. J.*, 42(3): 733–773, 1993.

[BV98] H. Bercovici and D. Voiculescu. Regularity questions for free convolution. In: Nonselfadjoint operator algebras, operator theory, and related topics, pp. 37–47, *Oper. Theory Adv. Appl.*, 104, Birkhäuser, Basel, 1998.

[CFK14] F. Cipriani, U. Franz, and A. Kula. Symmetries of Lévy processes on compact quantum groups, their Markov semigroups and potential theory. *J. Funct. Anal.* 266 (2014), 2789–2944.

[CG05] G.P. Chistyakov and F. Götze. The arithmetic of distributions in free probability theory. *Cent. Eur. J. Math.* 9 (2011), no. 5, 997–1050.

[CG06] G.P. Chistyakov and F. Götze. Limit theorems in free probability theory I. *Ann. Probab.* 36 (2008), no. 1, 54 –90.

[CGS96] F. Cipriani, D. Guido and S. Scarlatti, A remark on trace properties of K-cycles. *J. Operator Theory* 35 (1996), no. 1, 179–189.

[Cip97] F. Cipriani, Dirichlet forms and Markovian semigroups on standard forms of von Neumann algebras. *J. Funct. Anal.*, 147(2): 259–300, 1997.

[Cip08] F. Cipriani. Dirichlet forms on noncommutative spaces. In: Quantum Potential Theory, Lecture Notes in Math., Vol. 1954, pages 161–276, 2008.

[Con85] J.B. Conway. *A course in functional analysis.* Graduate Texts in Mathematics, 96. New York etc.: Springer-Verlag., 1985.

[Con94] A. Connes, *Noncommutative geometry.* Academic Press, Inc., San Diego, CA, 1994.

[CS03] F. Cipriani and J.L. Sauvageot, Derivations as square roots of Dirichlet forms. *J. Funct. Anal.*, 201(1): 78–120, 2003.

[CS15] M. Caspers and A. Skalski. The Haagerup approximation property for von Neumann algebras via quantum Markov semigroups and Dirichlet forms. *Comm. Math. Phys.* 336(3): 1637–1664, 2015.

[CU15] G. Cébron and M. Ulrich. Haar states and Lévy process on the unitary dual group. Work in progress. 2015.

[DK94] M.S. Dijkhuizen and T.H. Koornwinder. CQG algebras: a direct algebraic approach to compact quantum groups. *Lett. Math. Phys.* 32 (1994), no. 4, 315–330.

[DNR01] S. Dăscălescu, C. Năstăsescu and Ş. Raianu. *Hopf algebras. An introduction.* Monographs and Textbooks in Pure and Applied Mathematics, 235. Marcel Dekker, Inc., New York, 2001.

[FGT15] U. Franz, M. Gerhold, and A. Thom. Work in progress. 2015.

[FM05] U. Franz and N. Muraki. Markov structure of monotone Lévy processes. In H. Heyer, T. Hirai, T. Kawazoe, and K. Saito, editors, *Infinite Dimensional Harmonic Analysis III, Proceedings of the Third German-Japanese Symposium, University of Tübingen, Germany, 15–20 September 2003*, pages 37–57, 2005.

[Fra] U. Franz. Monotone increment processes and Loewner chains. In preparation.

[Fra00] U. Franz. Lévy processes on quantum groups. In *Probability on algebraic structures (Gainesville, FL, 1999)*, volume 261 of *Contemp. Math.*, pages 161–179. Amer. Math. Soc., Providence, RI, 2000.

[Fra01] U. Franz. Monotone independence is associative. *Infin. Dimens. Anal. Quantum Probab. Relat. Top.*, 4(3): 401–407, 2001.

[Fra02] U. Franz. What is stochastic independence? In: Non-commutativity, infinite-dimensionality and probability at the crossroads, 254–274, QP–PQ: Quantum Probab. White Noise Anal., 16, World Sci. Publ., River Edge, NJ, 2002.

[Fra03a] U. Franz. Lévy processes on real Lie algebras. First Sino-German Conference on Stochastic Analysis (A Satellite Conference of ICM 2002), Beijing, China, August 29 – September 3, 2002, 2003.

[Fra03b] U. Franz. Unification of boolean, monotone, anti-monotone, and tensor independence and Lévy process. *Math. Z.*, 243(4): 779–816, 2003.

[Fra03c] U. Franz. The theory of quantum Lévy processes. Habilitationsschrift, Ernst-Moritz-Arndt-Universität Greifswald, math.PR/0407488, 2003.

[Fra06a] U. Franz. Multiplicative monotone convolutions. *Banach Center Publications*, 73: 153–166, 2006.

[Fra06b] U. Franz. *Lévy Processes on Quantum Groups and Dual Groups.* In: [QIIP-II], pp. 16–257, 2006.

[Fra08] U. Franz. Boolean convolution of probability measures on the unit circle. In: Analyse et probabilités, pp. 83–94, Sémin. Congr., 16, Soc. Math. France, Paris, 2008.

[Fra09] U. Franz. Monotone and Boolean convolutions for non-compactly supported probability measures. *Indiana Univ. Math. J.* 58 (2009), no. 3, 1151–1185.

[Fre98] B. Fresse. Cogroups in algebras over an operad are free algebras. *Comment. Math. Helv.* 73 (1998) 637–676.

[FS99] U. Franz and R. Schott. *Stochastic Processes and Operator Calculus on Quantum Groups.* Kluwer Academic Publishers, Dordrecht, 1999.

[FS00] U. Franz and M. Schürmann. Lévy processes on quantum hypergroups. In *Infinite dimensional harmonic analysis (Kyoto, 1999)*, pages 93–114. Gräbner, Altendorf, 2000.

[FS08] U. Franz and A. Skalski, On ergodic properties of convolution operators associated with compact quantum groups. *Colloq. Math.*, 113(1): 13–23, 2008.

[Ger15] M. Gerhold. On Several Problems in the Theory of Comonoidal Systems and Subproduct Systems. Ph.D. Thesis, Ernst-Moritz-Arndt-Universität Greifswald, 2015.

[Gil98] R.D. Gill. Critique of 'Elements of quantum probability' [in *Quantum probability communications*, 73–100, World. Sci. Publ., River Edge, NJ] by B. Kümmerer and H. Maassen. In *Quantum Probability Communications*, pages 351–361. World Sci. Publishing, River Edge, NJ, 1998.

[GK13] O. Gühne and M. Kleinmann, Auf den Kontext kommt es an, *Physik Journal* 12, 25–30, 2013.

[GKS76] V. Gorini, A. Kossakowski, and E.C.G. Sudarshan. Completely positive dynamical semigroups of N-level systems. *J. Math. Phys.*, 17(5): 821–825, 1976.

[GL95] S. Goldstein and J.M. Lindsay, KMS-symmetric Markov semigroups. *Math. Z.*, 219(4): 591–608, 1995.

[GL14] M. Gerhold and S. Lachs, On (r, s)-products. Preprint, 2014.

[Gro55] A. Grothendieck, *Produits Tensoriel Topologiques et Espaces Nucleaires.* Memoires of the American mathematical Society. A.M.S. ed., Providence, Rhodes Island, 1955.

[GSS92] P. Glockner, M. Schürmann, and R. Speicher. Realization of free white noise. *Arch. Math.*, 58: 407–416, 1992.

[Gui72] A. Guichardet. *Symmetric Hilbert spaces and related topics*, volume 261 of *Lecture Notes in Math.* Springer-Verlag, Berlin, 1972.

[GvW89] P. Glockner and W. von Waldenfels. The relations of the noncommutative coefficient algebra of the unitary group. In *Quantum probability and applications, IV (Rome, 1987)*, volume 1396 of *Lecture Notes in Math.*, pages 182–220. Springer, Berlin, 1989.

[Hol01] A.S. Holevo. *Statistical structure of quantum theory*, volume 67 of *Lecture Notes in Physics. Monographs.* Springer-Verlag, Berlin, 2001.

[HP84] R.L. Hudson and K.R. Parthasarathy. Quantum Ito's formula and stochastic evolutions. *Comm. Math. Phys.*, 93(3): 301–323, 1984.

[HP86] R.L. Hudson and K.R. Parthasarathy. Unification of fermion and boson stochastic calculus. *Comm. Math. Phys.*, 104(3): 457–470, 1986.

[Hun56] G.A. Hunt. Semi-groups of measures on Lie groups. *Trans. Amer. Math. Soc.*, 81: 264–293, 1956.

[Koe91] H.T. Koelink, On $*$-representations of the Hopf $*$-algebra associated with the quantum group $\mathrm{U}_q(n)$. *Compositio Math.*, 77(2): 199–231, 1991.

[KK89] H.T. Koelink and T.H. Koornwinder, The Clebsch-Gordan coefficients for the quantum group $\mathrm{S}_\mu\mathrm{U}(2)$ and q-Hahn polynomials. *Nederl. Akad. Wetensch. Indag. Math.* 51 (1989), no. 4, 443–456.

[KS94] R. Koekoek and R.F. Swarttouw, The Askey-scheme of hypergeometric orthogonal polynomials and its q-analogue. Technical Report 94–05, Technical University of Delft, 1994. Available from math.CA/9602214.

[KS67] S. Kochen and E. Specker, The problem of hidden variables in quantum mechanics, *J. Math. Mech.* 17, 59–87, 1967.

[KS97] A. Klimyk and K. Schmüdgen, *Quantum groups and their representations*. Texts and Monographs in Physics. Springer-Verlag, Berlin, 1997.

[Kus97] J. Kustermans. Universal C^*-algebraic quantum groups arising from algebraic quantum groups. Preprint arXiv: funct-an/9704006, 1997.

[KV99] J. Kustermans and S. Vaes. A simple definition for locally compact quantum groups. *C. R. Acad. Sci. Paris Sér. I Math.* 328 (1999), no. 10, 871–876.

[KV00] J. Kustermans and S. Vaes. Locally compact quantum groups. *Ann. Sci. École Norm. Sup.* (4) 33 (2000), no. 6, 837–934.

[Kyp07] A.E. Kyprianou. *Introductory Lectures on Fluctuations of Lévy Processes with Applications.* Springer, 2007.

[Lac15] S. Lachs. A New Family of Universal Products and Aspects of a Non-Positive Quantum Probability Theory. Ph.D. Thesis, Ernst-Moritz-Arndt-Universität Greifswald, 2015.

[Len98] R. Lenczewski. Unification of independence in quantum probability. *Infin. Dimens. Anal. Quantum Probab. Relat. Top.*, 1(3): 383–405, 1998.

[Len01] R. Lenczewski. Filtered random variables, bialgebras, and convolutions. *J. Math. Phys.*, 42(12): 5876–5903, 2001.

[Len07] R. Lenczewski. Decompositions of the free additive convolution. *J. Funct. Anal.*, 246(2): 330–365, 2007.

[Li04] M. Liao, *Lévy processes in Lie groups. Cambridge Tracts in Mathematics*, vol. 162. Cambridge University Press, Cambridge, 2004.

[Lie99] V. Liebscher. On a central limit theorem for monotone noise. *Infin. Dimens. Anal. Quantum Probab. Relat. Top.*, 2(1): 155–167, 1999.

[Lin76] G. Lindblad. On the generators of quantum dynamical semigroups. *Comm. Math. Phys.*, 48(2): 119–130, 1976.

[Lin05] J.M. Lindsay. Quantum stochastic analysis – an introduction. In: [QIIP-I], pp. 181–271, 2005.

[LS05] J.M. Lindsay and A. Skalski, Quantum stochastic convolution cocycles. I. *Ann. Inst. H. Poincaré Probab. Statist.* 41 (2005), no. 3, 581–604.

[LS11] J.M. Lindsay and A. Skalski, Convolution semigroups of states. *Math. Z.*, 267(1–2) (2011), 325–339.

[LS12] J.M. Lindsay and A. Skalski, Quantum stochastic convolution cocycles III. *Math. Ann.* 352 (2012), no. 4, 779–804.

[Lu97] Y.G. Lu. An interacting free Fock space and the arcsine law. *Probab. Math. Statist.*, 17(1): 149–166, 1997.

[Maa92] H. Maassen. Addition of freely independent random variables. *J. Funct. Anal.*, 106(2): 409–438, 1992.

[Mac98] S. MacLane. *Categories for the working mathematician*, volume 5 of *Graduate texts in mathematics.* Springer-Verlag, Berlin, 2 edition, 1998.

[Maj95] S. Majid, *Foundations of quantum group theory.* Cambridge University Press, 1995.

[McC92] K. McClanahan. C^*-algebras generated by elements of a unitary matrix. *J. Funct. Anal.* 107 (1992), no. 2, 439–457.

[Mer85] N.D. Mermin. Is the moon there when nobody looks? Rreality and quantum theory. *Physics Today*, 38(4): 38–47, 1985.

[Mey95] P.-A. Meyer. *Quantum Probability for Probabilists*, volume 1538 of *Lecture Notes in Math.* Springer-Verlag, Berlin, 2nd edition, 1995.

[Mur97] N. Muraki. Noncommutative Brownian motion in monotone Fock space. *Comm. Math. Phys.*, 183(3): 557–570, 1997.

[Mur00] N. Muraki. Monotonic convolution and monotone Lévy-Hinčin formula. Preprint, 2000.

[Mur02] N. Muraki. The five independences as quasi-universal products. *Inf. Dim. Anal., quant. probab. and rel. fields*, 5(1): 113–134, 2002.

[Mur03] N. Muraki. The five independences as natural products. *Inf. Dim. Anal., quant. probab. and rel. fields*, 6(3): 337–371, 2003.

[Mur13] N. Muraki. A simple proof of the classification theorem for positive natural products. *Probab. Math. Stat.*, 33(2): 315–326, 2013.

[MT04] G.J. Murphy and L. Tuset. Aspects of compact quantum group theory, *Proc. Amer. Math. Soc.* 132 (2004), no. 10, 3055–3067.

[MvD98] A. Maes and A. Van Daele. Notes on Compact Quantum Groups, *Nieuw Arch. Wisk.* (4) 16 (1998), no. 1–2, 73–112.

[nlab Gl] `http://ncatlab.org/nlab/show/Gleason's+theorem`, Version of January 11, 2014.

[nlab KS] `http://ncatlab.org/nlab/show/Kochen-Specker+theorem`, Version of January 11, 2014.

[NS06] A. Nica and R. Speicher. *Lectures on the combinatorics of free probability*. London Mathematical Society Lecture Note Series, 335. Cambridge University Press, Cambridge, 2006.

[NT13] S. Neshveyev and L. Tuset. *Compact quantum groups and their representation categories.* Cours Spécialisés 20. Société Mathématique de France, Paris, 2013.

[Par90] K.R. Parthasarathy. Azéma martingales and quantum stochastic calculus. In R.R. Bahadur, editor, *Proc. R.C. Bose Memorial Symposium*, pages 551–569. Wiley Eastern, 1990.

[Par92] K.R. Parthasarathy. *An Introduction to Quantum Stochastic Calculus*. Birkhäuser, 1992.

[Par99] K.R. Parthasarathy. A Boson Fock space realization of arcsine Brownian motion. *Sankhyā Ser. A*, 61(3): 305–311, 1999.

[Par03] K.R. Parthasarathy. Lectures on quantum computation, quantum error-correcting codes and information theory, Notes by Amitava Bhattacharyya, TIFR, Mumbai, 2003.

[Pau02] V. Paulsen. *Completely bounded maps and operator algebras.* Cambridge University Press, Cambridge, 2002.

[Ped89] G.K. Pedersen. *Analysis now.* Graduate Texts in Mathematics, 118. New York etc.: Springer-Verlag., 1989.

[Pen89] R. Penrose. *The emperor's new mind.* The Clarendon Press Oxford University Press, New York, 1989. Concerning computers, minds, and the laws of physics, With a foreword by Martin Gardner.

[PS72] K.R. Parthasarathy and K. Schmidt. *Positive definite kernels, continuous tensor products, and central limit theorems of probability theory*, volume 272 of *Lecture Notes in Math.* Springer-Verlag, Berlin, 1972.

[PW90] P. Podleś and S.L. Woronowicz. Quantum deformation of Lorentz group. *Commun. Math. Phys.*, 130: 381–431, 1990.

[PW00] W. Pusz and S.L. Woronowicz, Representations of quantum Lorentz group on Gelfand spaces. *Rev. Math. Phys.* 12 (2000), no. 12, 1551Â1625.

[QIIP-I] D. Applebaum, B.V.R. Bhat, J. Kustermans and J.M. Lindsay. *Quantum Independent Increment Processes I: From Classical Probability to Quantum Stochastic Calculus* U. Franz, M. Schürmann (eds.), Lecture Notes in Math., Vol. 1865, Springer, 2005.

[QIIP-II] O.E. Barndorff-Nielsen, U. Franz, R. Gohm, B. Kümmerer and S. Thorbjørnsen. *Quantum Independent Increment Processes II: Structure of Quantum Lévy Processes, Classical Probability and Physics*, U. Franz, M. Schürmann (eds.), Lecture Notes in Math., Vol. 1866, Springer, 2006.

[RS80] M. Reed and B. Simon. *Methods of modern mathematical physics. I.* Academic Press Inc. [Harcourt Brace Jovanovich Publishers], New York, second edition, 1980. Functional analysis.

[RW00a] L.C.G. Rogers and D. Williams. *Diffusions, Markov processes, and martingales. Vol. 1. Foundations.* Reprint of the second (1994) edition. Cambridge University Press, Cambridge, 2000.

[RW00b] L.C.G. Rogers and D. Williams. *Diffusions, Markov processes, and martingales. Vol. 2. Itô calculus.* Reprint of the second (1994) edition. Cambridge University Press, Cambridge, 2000.

[Sat99] K.-I. Sato. *Lévy Processes and Infinitely Divisible Distributions.* Cambridge University Press, Cambridge, 1999.

[Sch90a] M. Schürmann. Gaussian states on bialgebras. In: Quantum probability and applications, V (Heidelberg, 1988), 347–367, Lecture Notes in Math., 1442, Springer, Berlin, 1990.

[Sch90b] M. Schürmann. Noncommutative stochastic processes with independent and stationary increments satisfy quantum stochastic differential equations. *Probab. Theory Related Fields*, 84(4): 473–490, 1990.

[Sch91a] M. Schürmann. The Azéma martingales as components of quantum independent increment processes. In J. Azéma, P.A. Meyer, and M. Yor, editors, *Séminaire de Probabilités XXV*, volume 1485 of *Lecture Notes in Math.* Springer-Verlag, 1991.

[Sch91b] M. Schürmann. Quantum stochastic processes with independent additive increments. *J. Multivariate Anal.*, 38(1): 15–35, 1991.

[Sch93] M. Schürmann. *White Noise on Bialgebras*, volume 1544 of *Lecture Notes in Math.* Springer-Verlag, Berlin, 1993.

[Sch95a] M. Schürmann. Direct sums of tensor products and noncommutative independence. *J. Funct. Anal.*, 1995.

[Sch95b] M. Schürmann. Noncommutative probability on algebraic structures. In H. Heyer, editor, *Proceedings of XI Oberwolfach Conference on Probability Measures on Groups and Related Structures*, pages 332–356. World Scientific, 1995.

[Sch97] M. Schürmann. Cours de DEA. Université Louis Pasteur, Strasbourg, 1997.

[Sch00] M. Schürmann. Operator processes majorizing their quadratic variation. *Infin. Dimens. Anal. Quantum Probab. Relat. Top*, 3(1): 99–120, 2000.

[Ske99] M. Skeide. Hunt's Formula for $SU_q(2)$ – a Unified View. *Open Systems & Information Dynamics*, 6(1): 1–28, 1999.

[Ske01] M. Skeide. Hilbert modules and applications in quantum probability. Habilitation thesis, 2001.

[Sko91] A.V. Skorohod. *Random Processes with Independent Increments*. Kluwer Academic Publishers, 1991. (Russian original 1986).

[Spe97] R. Speicher. Universal products. In D. Voiculescu, editor, *Free probability theory. Papers from a workshop on random matrices and operator algebra free products, Toronto, Canada, March 1995*, volume 12 of *Fields Inst. Commun.*, pages 257–266. American Mathematical Society, Providence, RI, 1997.

[SS98] M. Schürmann and M. Skeide, Infinitesimal generators of the quantum group $SU_q(2)$. *Inf. Dim. Anal., Quantum Prob. and Rel. Topics*, 1(4): 573–598, 1998.

[SSV10] M. Schürmann, M. Skeide, and S. Volkwardt. Transformation of quantum Lévy processes on Hopf algebras. *Commun. Stoch. Anal.*, 4(4): 553–577, 2010.

[SV08] M. Schürmann and S. Voß. Positivity of free convolution semigroups. In: *Quantum probability and related topics*, volume 23 of QP–PQ: Quantum Probab. White Noise Anal., pages 225–235. World Sci. Publ., Hackensack, NJ, 2008.

[SV14] M. Schürmann and S. Voß. Schoenberg Correspondence on Dual Groups. *Comm. Math. Phys.* 328 (2014), no. 2, 849–865.

[Str00] R.F. Streater. Classical and quantum probability. *J. Math. Phys.*, 41(6): 3556–3603, 2000.

[SW97] R. Speicher and R. Woroudi. Boolean convolution. In D. Voiculescu, editor, *Free probability theory. Papers from a workshop on random matrices and operator algebra free products, Toronto, Canada, March 1995*, volume 12 of *Fields Inst. Commun.*, pages 267–279. American Mathematical Society, Providence, RI, 1997.

[Swe69] M.E. Sweedler. *Hopf Algebras*. Benjamin, New York, 1969.

[Tim08] T. Timmermann. *An invitation to quantum groups and duality. From Hopf algebras to multiplicative unitaries and beyond.*

EMS Textbooks in Mathematics. European Mathematical Society (EMS), Zürich, 2008.

[Ulr14] M. Ulrich. Construction of a free Lévy process as high-dimensional limit of a Brownian motion on the Unitary group. Preprint arXiv: 1407.0212v1, 2014. To appear in Inf. Dim. Anal., Quantum Prob. and Rel. Topics.

[VD97] A. Van Daele, The Haar measure on finite quantum groups. *Proc. Amer. Math. Soc.*, 125(12): 3489–3500, 1997.

[VD98] A. Van Daele. An algebraic framework for group duality. *Adv. Math.*, 140(2): 323–366, 1998.

[VD03] A. Van Daele. Multiplier Hopf*-algebras with positive integrals: a laboratory for locally compact quantum groups. Vainerman, Leonid (ed.), Locally compact quantum groups and groupoids. Proceedings of the 69th meeting of theoretical physicists and mathematicians, Strasbourg, France, February 21–23, 2002. Berlin: Walter de Gruyter. IRMA Lect. Math. Theor. Phys. 2, 229–247 (2003)., 2003.

[VDN92] D. Voiculescu, K. Dykema, and A. Nica. *Free Random Variables*. AMS, 1992.

[VDW96] A. Van Daele and Shuzhou Wang. Universal quantum groups. *Internat. J. Math.*, 7(2): 255–263, 1996.

[Voi86] D. Voiculescu. Addition of certain noncommuting random variables. *J. Funct. Anal.*, 66(3): 323–346, 1986.

[Voi87] D. Voiculescu. Dual algebraic structures on operator algebras related to free products. *J. Oper. Theory*, 17: 85–98, 1987.

[Voi90] D. Voiculescu. Noncommutative random variables and spectral problems in free product C^*-algebras. *Rocky Mountain J. Math.*, 20(2): 263–283, 1990.

[Voi91] D. Voiculescu. Limit laws for random matrices and free products. *Invent. Math.* 104 (1991), no. 1, 201–220.

[Voi00] D. Voiculescu. Lectures on free probability theory. In: *Lectures on probability theory and statistics (Saint-Flour, 1998)*, pp. 279–349, Lecture Notes in Math., 1738, Springer, Berlin, 2000.

[Voi11] C. Voigt. The Baum-Connes conjecture for free orthogonal quantum groups. *Adv. Math.*, 227(5): 1873–1913, 2011.

[Vos13] S. Voß. Realisierung von Quantum-Lévy-Prozessen auf Fockräumen. Ph.D. Thesis, Ernst-Moritz-Arndt-Universität Greifswald, 2013.

[Wal73] W. von Waldenfels. An approach to the theory of pressure broadening of spectral lines. In M. Behara, K. Krickeberg, and J. Wolfowitz, editors, *Probability and Information Theory II*, volume 296 of *Lecture Notes in Math.* Springer–Verlag, Berlin, 1973.

[Wal84] W. von Waldenfels. Ito solution of the linear quantum stochastic differential equation describing light emission and absorption. In *Quantum probability and applications to the quantum theory of irreversible processes, Proc. int. Workshop, Villa Mondragone/Italy 1982*, volume 1055 of *Lecture Notes in Math.*, pages 384–411. Springer–Verlag, 1984.

[wiki Bell] `http://en.wikipedia.org/wiki/Bell's_theorem`, Version of July 15, 2014.

[wiki bra-ket] `http://en.wikipedia.org/wiki/Bra-ket_notation`, Version of July 15, 2014.

[Wor87a] S.L. Woronowicz. Compact matrix pseudogroups. *Commun. Math. Phys.*, 111: 613–665, 1987.

[Wor87b] S.L. Woronowicz, Twisted $\mathrm{SU}(2)$ group. An example of a noncommutative differential calculus. *Publ. Res. Inst. Math. Sci.*, 23(1): 117–181, 1987.

[Wor88] S.L. Woronowicz, Tannaka-Kreĭn duality for compact matrix pseudogroups. Twisted $\mathrm{SU}(N)$ groups. *Invent. Math.*, 93(1): 35–76, 1988.

[Wor98] S.L. Woronowicz. Compact quantum groups. In A. Connes, K. Gawedzki, and J. Zinn-Justin, editors, *Symétries Quantiques, Les Houches, Session LXIV, 1995*, pages 845–884. Elsevier Science, 1998.

[Zha91] J.J. Zhang. H-algebras. *Adv. Math.*, 89(2): 144–191, 1991.

2

Quantum Dynamical Systems from the Point of View of Noncommutative Mathematics

A starting point of 'noncommutative mathematics' is based on the observation that many features of classical spaces can be phrased in terms of the algebras of functions defined on these spaces. In these lectures we will present a few examples of how in presence of dynamics this idea leads to the study of 'quantum' transformations. Our focus will be on two building blocks of modern abstract theory of dynamical systems: entropy and ergodic theorems.

2.1 Noncommutative Mathematics and Quantum/ Noncommutative Dynamical Systems

The notion of quantum processes, or quantum dynamical systems, can be understood in several different ways, as can be viewed for example by looking at some of the several hundred articles classified in the MathSciNet database under the heading MSC 46L53 (corresponding to 'Noncommutative dynamical systems'). In these lectures we will present a purely mathematically motivated approach to certain 'quantum' or 'noncommutative' versions of fundamental concepts and problems studied in classical topological and measurable dynamics. The origins of the need to study questions of that type can be naturally found in quantum mechanics: readers interested in physical interpretations of the topics studied below can consult, for example, the book [AF], or the lecture of Uwe Franz in the first part of this volume. Here our approach will be based on what is nowadays often called the philosophy of 'Noncommutative Mathematics'.

Notational conventions

For a subset X of a Banach space the closed linear span of X will be denoted by $\overline{\text{Lin}}\, X$ (and the usual linear span by Lin X). Notation $F \subset\subset Z$ will be used to signify that F is a finite subset of Z. Sometimes we will write $\mathbb{N}_0$ for $\mathbb{N} \cup \{0\}$. All vector spaces (algebras, Hilbert spaces, and so on) will be considered over $\mathbb{C}$. As in the first part of this volume, scalar products are linear on the *right* (and not on the wrong!) side.

2.1.1 Noncommutative Mathematics – Gelfand–Naimark Theorem

Definition 2.1.1 A Banach algebra A with involution (that is, a complex algebra with involutive antilinear antimultiplicative map $^* : \mathsf{A} \to \mathsf{A}$, equipped with a submultiplicative norm making A a Banach space and such that the involution is an isometry) is called a C^*-algebra if

$$\|a^*a\| = \|a\|^2, \quad a \in \mathsf{A}.$$

For proofs of various general statements related to C^*-algebras that will be used below we refer, for example, to the monograph [Mu]. In these lectures **all C^*-algebras will be unital**. A basic example of a C^*-algebra is M_n, the algebra of n by n matrices with complex coefficients, equipped with the natural involution and multiplication and with the operator norm induced by the identification of each such matrix with a linear operator acting on the Hilbert space $\mathbb{C}^n$. More generally, whenever H is a Hilbert space, $B(\mathsf{H})$, that is, the algebra of all bounded linear operators on H, is a C^*-algebra. Another example is given by $C(X)$, the algebra of all continuous complex-valued functions on a compact space X, equipped with the natural algebraic operations and the supremum norm (we will assume that all compact spaces are Hausdorff).

Theorem 2.1.2 [Gelfand-Naimark, 1943]: Every commutative unital C^*-algebra A is isometrically isomorphic to the algebra $C(X_{\mathsf{A}})$ for some compact topological space X_{A}. If Y is another compact space then the C^*-algebras A and $C(Y)$ are isometrically isomorphic if and only if X_{A} and Y are homeomorphic.

The space X_{A} is called the *spectrum* of the commutative algebra A.

The assumption of unitality of the C^*-algebras that we study is in a sense equivalent to assuming the compactness of the underlying (quantum) space; for example each commutative C^*-algebra is isomorphic to $C_0(X)$, the algebra of continuous functions vanishing at infinity on a locally compact space X, and of course $C_0(X)$ is unital if and only if X is compact. In general, the procedure of adding a unit (unitisation) can be viewed as a noncommutative counterpart of passing to a compactification of a given locally compact space.

Exercise 2.1.1 Let H be a Hilbert space and let $\mathcal{K}_{\mathsf{H}}$ denote the space of all compact operators on H; recall it is the norm closure of the space of all finite-rank operators on H. Show that $\mathcal{K}_{\mathsf{H}}$ is a C^*-subalgebra of $B(\mathsf{H})$, unital if and only if H is finite-dimensional. Furthermore, show that $\mathcal{K}_{\mathsf{H}} + \mathbb{C}1 := \{k + \lambda 1_{B(\mathsf{H})} : k \in \mathcal{K}_{\mathsf{H}}, \lambda \in \mathbb{C}\}$ is a unital C^*-subalgebra of $B(\mathsf{H})$.

2.1.2 Quantum topological dynamical systems and some properties of transformations of C^*-algebras

Definition 2.1.3 A (classical) topological dynamical system is a pair (X, T), where X is a compact space and $T : X \to X$ a continuous map.

If X, Y are compact spaces and $T : X \to Y$ is continuous, then we obtain in a natural way a map $\alpha_T : C(Y) \to C(X)$ defined by the formula:

$$\alpha_T(f) = f \circ T, \quad f \in C(Y). \tag{2.1.1}$$

Note that 'the arrows are inverted':

$$\begin{array}{llcll} T: & X & \to & Y & \\ & C(X) & \leftarrow & C(Y) & : \alpha_T. \end{array}$$

It is easy to check that α_T is a unital *-homomorphism. It turns out that every unital *-homomorphism $\alpha : C(Y) \to C(X)$ arises in this way from exactly one continuous transformation from X to Y. Thus, the 'classical' theory of compact spaces and continuous maps is in a natural way equivalent to the theory of commutative unital C^*-algebras and unital *-homomorphisms.

Definition 2.1.4 A quantum (or noncommutative) dynamical system is a pair (A, α), where A is a unital C^*-algebra and $\alpha : \mathsf{A} \to \mathsf{A}$ a unital *-homomorphism.

In these lectures we will study mainly discrete dynamics, described by a single transformation. As in the classical setting one can also investigate dynamical systems described by discrete or continuous group actions (for example, families of automorphisms of a given C^*-algebra indexed by $\mathbb{R}_+$ or $\mathbb{R}$). In quantum mechanics traditionally the most important role was played by automorphisms of $B(\mathsf{H})$, which are automatically *inner*, that is, given by the formula $a \mapsto U^*aU$, where $U \in B(\mathsf{H})$ is a *unitary* ($UU^* = U^*U = 1$) – see also Section 2.4.1 below.

An element of a C^*-algebra A is called *positive* if it is equal to b^*b for some $b \in \mathsf{A}$. The collection of all positive elements in A, denoted A_+, is a closed cone in A; moreover, if $a, b \in \mathsf{A}_+$ then $\|a + b\| \geq \|a\|$. A linear map between C^*-algebras is called *positive* if it maps positive elements to positive elements. In particular, we can talk about positive functionals; it is easy to show that a positive functional is automatically bounded.

Definition 2.1.5 A state on a C^*-algebra A is a positive linear functional $\omega : \mathsf{A} \to \mathbb{C}$ such that $\omega(1) = 1$.

The collection of all states on a given C^*-algebra A will be denoted by $S(\mathsf{A})$; a state $\omega \in S(\mathsf{A})$ is said to be *faithful* if for a fixed $a \in \mathsf{A}$ the equality $\omega(a^*a) = 0$ implies that $a = 0$ and *tracial* if $\omega(ab) = \omega(ba)$ for all $a, b \in \mathsf{A}$. States have the role of quantum probability measures – Riesz theorem says that if X is a compact space then there exists a bijective correspondence between states on $C(X)$ and Borel regular probability measures on X (the value of the state given by a measure μ on a function $f \in C(X)$ is given by the integration of f with respect to μ). Thus, tracial states (also known simply as *traces*) are those which preserve vestiges of commutativity: informally speaking, when we work with traces on noncommutative algebras we think of 'functions' that 'commute under the integral sign'.

A unital *-homomorphism from a C^*-algebra A to $B(\mathsf{H})$ is called a *representation* of A. We say that a representation $\pi : \mathsf{A} \to B(\mathsf{H})$ is *faithful* if it is injective (in that case it is automatically isometric; more generally each *-homomorphism between C^*-algebras is contractive). Another theorem due to Gelfand and Naimark states that each C^*-algebra has a faithful representation — in other words, each C^*-algebra is isomorphic to a closed *-subalgebra of $B(\mathsf{H})$ for certain Hilbert space H.

Applying the last statement one can easily show that if A is a C^*-algebra, then the *-algebra $M_n(\mathsf{A})$ of n by n matrices with coefficients in A possesses a natural (and unique) C^*-norm. This turns out to have a very important role in the theory. If A, B, are C^*-algebras and $T : \mathsf{A} \to \mathsf{B}$ is a linear map, then applying T separately to each entry of the matrix yields a linear map $T^{(n)} : M_n(\mathsf{A}) \to M_n(\mathsf{B})$. We say that T is *completely positive,* if $T^{(n)}$ is positive for each $n \in \mathbb{N}$. Sometimes it is convenient to identify $M_n(\mathsf{A})$ with the algebraic tensor product $M_n \otimes \mathsf{A}$. If $T : \mathsf{A} \to \mathsf{A}$, then the map $T^{(n)} : M_n(\mathsf{A}) \to M_n(\mathsf{A})$ corresponds under this identification to the map $\mathrm{id}_{M_n} \otimes T : M_n \otimes \mathsf{A} \to M_n \otimes \mathsf{A}$.

Remark 2.1.6 Each *-homomorphism acting between C^*-algebras is completely positive. If A or B is commutative and $T : \mathsf{A} \to \mathsf{B}$ is a positive map, then T is automatically completely positive; therefore, in particular states are completely positive. Further, sums and compositions of completely positive maps are completely positive and, if $\mathsf{A} \subset B(\mathsf{H})$ and $V \in B(\mathsf{H}; \mathsf{K})$, where H, K are Hilbert spaces, then the map $a \mapsto V^*aV$ is a completely positive map from A to $B(\mathsf{K})$. All unital completely positive maps are contractive.

The following theorem is sometimes called the 'quantum/noncommutative Hahn–Banach theorem'.

Theorem 2.1.7 [Arveson Extension Theorem]: Let H be a Hilbert space, A – a C^*-algebra, and B – a C^*-subalgebra of A. If $T : \mathsf{B} \to B(\mathsf{H})$ is completely positive, then there exists a completely positive map $\widetilde{T} : \mathsf{A} \to B(\mathsf{H})$ such that $\widetilde{T}|_{\mathsf{B}} = T$ and $\|T\| = \|\widetilde{T}\|$.

Finally, note that if we want to study 'noncommutative counterparts' of compact spaces equipped with some further structure, say for example *compact groups,* we can do it, provided we find a way of encoding of this extra structure solely in terms of the algebra of continuous functions defined on the space and possibly some transformations acting on algebras of that type. In the particular example listed above such an approach was successfully implemented by Woronowicz ([Wo]) and led to the theory of *compact quantum groups*, which is of fundamental importance to the lectures of Uwe Franz in the first part of this volume.

2.1.3 Cuntz algebras

Examples of C^*-algebras are usually defined either *abstractly*, as completions of certain *-algebras with respect to a given norm satisfying the C^*-condition, or *concretely*, as specific closed *-subalgebras of $B(\mathsf{H})$. Below we present a very important example of an abstractly defined C^*-algebra.

Definition 2.1.8 ([Cu$_1$]) Let $N \in \mathbb{N}, N \geq 2$. Consider a unital *-algebra $\mathcal{A}_N$ generated by N elements $S_1, \ldots, S_N$, which satisfy the following relations:

$$S_i^* S_j = \delta_{i,j} 1_{\mathcal{A}_N}, \quad \sum_{i=1}^{n} S_i S_i^* = 1_{\mathcal{A}_N}.$$

The Cuntz algebra $\mathcal{O}_N$ is the completion of $\mathcal{A}_N$ in the norm

$$\|x\| = \sup\{\|\pi(x)\| : \pi\text{−unital } ^*\text{-homomorphism from } \mathcal{A}_N \text{ to } B(\mathsf{H})\}.$$

In other words, the Cuntz algebra is the universal C^*-algebra generated by N isometries with orthogonal ranges, whose range projections sum up to identity. Note that the fact that the above formula defines a norm on $\mathcal{A}_N$ (and not only a seminorm) requires a proof.

Let $\mathcal{J}_k = \{1, \ldots, N\}^k$ denote the set of all multi-indices of length k with values in $\{1, \ldots, N\}$. For $\mu \in \mathcal{J}_k$ put

$$S_\mu := S_{\mu_1} S_{\mu_2} \ldots S_{\mu_k},$$
$$|\mu| = k.$$

Theorem 2.1.9 ([Cu$_1$]) The algebra $\mathcal{O}_N$ is *simple*, that is, does not possess any non-trivial two-sided closed ideals. In particular any C^*-algebra generated by N-isometries acting on a given Hilbert space H and whose ranges are mutually orthogonal and together fill H is isomorphic to $\mathcal{O}_N$. The set

$$\text{Lin}\{S_\mu S_\nu^* : \mu \in \mathcal{J}_k, \nu \in \mathcal{J}_l, k, l \in \mathbb{N}\}$$

is a dense *-subalgebra of $\mathcal{O}_N$.

We will be further interested in the following two C^*-subalgebras of $\mathcal{O}_N$:

$$\mathcal{F}_N = \overline{\mathrm{Lin}}\,\{S_\mu S_\nu^* : \mu \in \mathcal{J}_k, \nu \in \mathcal{J}_k, k \in \mathbb{N}\},$$

$$\mathcal{C}_N = \overline{\mathrm{Lin}}\,\{S_\mu S_\mu^* : \mu \in \mathcal{J}_k, k \in \mathbb{N}\}.$$

Exercise 2.1.2 Prove that $\mathcal{C}_N$ is an abelian subalgebra of $\mathcal{O}_N$ (in fact it is even a *masa* – a maximal abelian subalgebra).

In connection with $\mathcal{C}_N$ consider now the Cantor set, presented in the form given by a collection of infinite words built of letters from the alphabet $\{1, \ldots, N\}$:

$$\mathfrak{C}_N = \{w = (w_k)_{k=1}^\infty : \forall_{k \in \mathbb{N}}\, w_k \in \{1, \ldots, N\}\}.$$

The set $\mathfrak{C}_N$ is equipped with a natural metric

$$d(w, v) = \begin{cases} 0 & \text{if} \quad w = v, \\ \frac{1}{k} & \text{if} \quad w_k \neq v_k,\ w_i = v_i \text{ for } i < k. \end{cases}$$

It is easy to check that $(\mathfrak{C}_N, d)$ is a compact metric space (and the resulting topology coincides with the Tikhonov topology of the infinite Cartesian product of the discrete sets $\{1, \ldots, N\}$).

Theorem 2.1.10 The map sending each $S_\mu S_\mu^*$ ($\mu \in \bigcup_{k \in \mathbb{N}} \mathcal{J}_k$) to the characteristic function of the set $Z_\mu := \{w \in \mathfrak{C}_N : (w_n)_{n=1}^{|\mu|} = \mu\}$ extends to a *-isomorphism between $\mathcal{C}_N$ and $C(\mathfrak{C}_N)$, which will be denoted in what follows by γ.

Proof Exercise. □

We can thus think of $\mathcal{O}_N$ as a certain noncommutative C^*-algebra containing the algebra of continuous functions on the Cantor set as a subalgebra.

Exercise 2.1.3 Let X, Y be compact spaces and assume that we have a natural (unital, injective, *-homomorphic) inclusion of $C(X)$ into $C(Y)$ (in other words, $C(X)$ is a unital subalgebra of $C(Y)$). What can be then said about the sets X and Y? On the other hand, suppose that X is a closed subset of a compact space Y. What can be then said of algebras $C(X)$ and $C(Y)$?

Exercise 2.1.4 Show that $\text{Lin}\{S_\mu S_\nu^* : \mu \in \mathcal{J}_k, \nu \in \mathcal{J}_k, k \in \mathbb{N}\}$ is the union of an increasing sequence of finite-dimensional unital *-algebras $\mathcal{F}_N^l = \text{Lin}\{S_\mu S_\nu^* : \mu \in \mathcal{J}_k, \nu \in \mathcal{J}_k, k \leq l\}$ $(l \in \mathbb{N})$. Identify $\mathcal{F}_N^l$ with a familiar (by now) C^*-algebra and interpret $\text{Lin}\{S_\mu S_\mu^* : \mu \in \mathcal{J}_k, k \leq l\}$ as a subalgebra of $\mathcal{F}_N^l$.

2.1.4 Graph C^*-algebras

Cuntz algebras turn out to be special examples of a very important and wide class of C^*-algebras associated to (finite) directed graphs. They first appeared in literature as generalizations of Cuntz–Krieger algebras (to be described in what follows), which in turn generalize Cuntz algebras introduced above.

Let $\Lambda = (V, E, r, s)$ be a *finite directed graph*. By that we mean that V and E are finite sets (thought of, respectively, as sets of vertices and edges of Λ) and $r, s : E \to V$ certain functions (thought of as attributing to each edge its source and range vertex). A (finite) *path* in Λ is a finite sequence of edges, say $\nu := (e_i)_{i=1}^k$ such that for all $i = 1, \cdots, k-1$ we have $s(e_i) = r(e_{i+1})$. The number k is called the length of the path ν (which we write as $|\mu| = k$); the path ν has the *range* $r(\nu) := r(e_1)$ and the *source* $s(\nu) := s(e_k)$. Thus, edges can be viewed as paths of length 1; moreover, we can view vertices as paths of length 0. Compatible paths can be concatenated: if μ and ν are paths, and $s(\mu) = r(\nu)$, then the path $\mu\nu$ is given by consecutive enumeration of edges belonging to μ followed by the enumeration of edges belonging to ν. The set of all finite paths (including paths of length 0) will be denoted by $\mathcal{P}_\Lambda$. We say that Λ *contains no sources* if for each $v \in V$ and each $m \in \mathbb{N}$ there exists a path μ of length m such that $r(\mu) = v$. Generalizing the idea of paths leads to the notion of the space of infinite paths, $\mathcal{P}_\Lambda^\infty$: these are functions from $\mathbb{N}_0$ to E such that for all $i \in \mathbb{N}_0$ the source of the i-th edge is equal to the range of the $i+1$-th edge. Note that if the set of edges has cardinality N, then $\mathcal{P}_\Lambda^\infty$ is a (closed) subset of the Cantor set $\mathfrak{C}_N$; moreover, $\mathcal{P}_\Lambda^\infty$ is non-empty if Λ contains no sources.

A collection of operators $\{P_v : v \in V\} \cup \{S_e : e \in E\}$ in some C^*-algebra is called a *non-degenerate Cuntz–Krieger family* (associated with Λ) if the following conditions hold ($v \in V, e \in E$):

(i) each P_v is an orthogonal projection, that is, $P_v = P_v{}^* = P_v{}^2$;

(ii) $S_e{}^* S_e = P_{s(e)}$ (in particular each S_e is a *partial isometry*, that is $S_e S_e{}^* S_e = S_e$);

(iii) $P_v = \sum_{f:r(f)=v} S_f S_f{}^*$;

(iv) $\sum_{w \in V} P_w = 1$.

Note that the relations above imply, for example, that if $e, f \in E$ and $e \neq f$, then $S_e{}^* S_f = 0$ (consider separately the cases when $r(e) = r(f)$ and $r(e) \neq r(f)$). Similarly, one can show that $S_e S_f = 0$ if $e, f \in E, r(f) \neq s(e)$. This suggests that we can usefully define for a given Cuntz–Krieger family also partial isometries corresponding to paths in $\mathcal{P}_\Lambda$: if $\nu := (e_i)_{i=1}^k$ is a path, we put $S_\nu = S_{e_1} \cdots S_{e_k}$.

Definition 2.1.11 The graph C^*-algebra $C^*(\Lambda)$ associated to a finite directed graph Λ with no sources is the universal C^*-algebra generated by a non-degenerate Cuntz–Krieger family.

The above definition has to be understood precisely in the way in which the Cuntz algebra was introduced in Definition 2.1.8. In particular, $C^*(\Lambda)$ is equipped with a canonical non-degenerate Cuntz–Krieger family $\{P_v : v \in V\} \cup \{S_e : e \in E\}$; moreover, using the commutation relations satisfied by Cuntz–Krieger families one can show that the set $\{S_\nu S_\mu^* \; : \; \mu, \nu \in \mathcal{P}_\Lambda\}$ is linearly dense in $C^*(\Lambda)$.

Exercise 2.1.5 Show that if $N \geq 2$ and Λ is a graph with a single vertex and N edges, then $C^*(\Lambda) \approx \mathcal{O}_N$. Analyse the connections between paths and multi-indices.

Exercise 2.1.6 Experiment with some graphs having two vertices that contain no sources and see if you can identify the associated graph C^*-algebras.

Another important class of graph C^*-algebras is determined by $\{0,1\}$-valued matrices: if $k \in \mathbb{N}$ and $M \in M_k(\{0,1\})$ then the asssociated graph Λ_M is given by the set of k vertices and the collection of edges determined by the condition $(i, j) \in E$ if and only if $M(i, j) = 1$. The resulting graph C^*-algebra is called the *Cuntz–Krieger algebra* associated to M.

As in the case of Cuntz algebras, we will be interested in what follows in some subalgebras of $C^*(\Lambda)$. In particular, define

$$\mathcal{F}_\Lambda = \overline{\text{Lin}}\,\{S_\mu S_\nu^* : \mu, \nu \in \mathcal{P}_\Lambda,\ |\mu| = |\nu|\},$$

$$\mathcal{C}_\Lambda = \overline{\text{Lin}}\,\{S_\mu S_\mu^* : \mu \in \mathcal{P}_\Lambda\}.$$

The second of the algebras above can be shown to be commutative; in fact, following arguments similar to those of Theorem 2.1.10 one can prove that it is isomorphic to the algebra of continuous functions on $\mathcal{P}_\Lambda^\infty$. We will denote the corresponding isomorphism from $\mathcal{C}_\Lambda$ to $C(\mathcal{P}_\Lambda^\infty)$ by γ_Λ. The structure of $\mathcal{C}_\Lambda$ is slightly more complicated to analyze — it is however always an *AF-algebra,* that is, an inductive limit of a sequence of finite direct sums of matrix algebras.

The construction described in this section has far-reaching extensions to the case of infinite graphs, to *higher-rank graphs* of Kumjian and Pask, which can be viewed as colored graphs equipped with special rules identifying certain paths built of edges of different colors, or to *topological graphs.* A rich and accessible source of information about various graph C^*-algebras can be found in the monograph [Ra].

2.1.5 C^*-algebras associated with discrete groups

In the following sections we will use a few times examples based on certain C^*-algebras associated with discrete groups. Here we introduce the relevant construction.

Let Γ be a discrete group and consider the Hilbert space $\ell^2(\Gamma)$. Each element $\gamma \in \Gamma$ defines a *left shift* on $\ell^2(\Gamma)$, which is a bounded operator $\lambda_\gamma \in B(\ell^2(\Gamma))$, by the formula

$$((\lambda_\gamma)(f))(\gamma') = f(\gamma^{-1}\gamma'), \quad f \in \ell^2(\Gamma), \gamma' \in \Gamma.$$

It is easier to understand the definition of λ_γ if we write how it acts on the elements of the type $\delta_{\gamma'} \in \ell^2(\Gamma)$, which form an orthonormal basis in $\ell^2(\Gamma)$:

$$\lambda_\gamma \delta_{\gamma'} = \delta_{\gamma\gamma'}, \quad \gamma, \gamma' \in \Gamma. \tag{2.1.2}$$

Then it is easy to see that the collection of operators $(\lambda_\gamma)_{\gamma\in\Gamma}$ satisfies the following properties:

(i) each λ_γ is unitary;

(ii) for any $\gamma, \gamma' \in \Gamma$ we have

$$\lambda_\gamma \lambda_{\gamma'} = \lambda_{\gamma\gamma'}.$$

Definition 2.1.12 The C^*-algebra generated in $B(\ell^2(\Gamma))$ by the collection of operators $\{\lambda_\gamma\}_{\gamma \in \Gamma}$ is called the reduced C^*-algebra of Γ and denoted by $C_r^*(\Gamma)$.

Note that the Cuntz algebras and graph C^*-algebras introduced above were defined as *abstract* C^*-algebras, that is, completions of some *-algebras with respect to the universal C^*-norm. The algebra $C_r^*(\Gamma)$ is defined as a *concrete* C^*-algebra, that is, as a specific closed unital *-subalgebra of $B(\mathsf{H})$ for a specific Hilbert space H. It is possible to associate with Γ also a natural abstract C^*-algebra, the *universal (or full) C^*-algebra of* Γ . It is denoted simply as $C^*(\Gamma)$ and arises as the universal C^*-completion of the purely algebraic group ring $\mathbb{C}[\Gamma]$. As $C_r^*(\Gamma)$ can be viewed as *a* C^*-completion of $\mathbb{C}[\Gamma]$, the definitions imply the existence of the canonical surjective unital *-homomorphism $\pi : C^*(\Gamma) \to C_r^*(\Gamma)$. Note, however, that π need not be injective (it is injective if and only if Γ is amenable, see Section 2.4).

When Γ is abelian, then $C_r^*(\Gamma)$ is commutative and isomorphic to the C^*-algebra $C(\hat{\Gamma})$, where $\hat{\Gamma}$ is a compact space, the *dual group* (the *Pontriagin dual*) of Γ. This is essentially a Fourier transform fact; note that it can be viewed as a starting point to introduce the noncommutative objects 'dual' to classical, not necessarily abelian, discrete groups.

Exercise 2.1.7 Decribe explicitly the isomorphism between $C_r^*(\Gamma)$ and $C(\hat{\Gamma})$ whose existence is stated above for finite cyclic groups ($\mathbb{Z}/_{n\mathbb{Z}}, n \in \mathbb{N}$). Remember that $\widehat{\mathbb{Z}/_{n\mathbb{Z}}} = \mathbb{Z}/_{n\mathbb{Z}}$.

2.2 Noncommutative Topological Entropy of Voiculescu

In this section, we introduce certain classes of quantum dynamical systems illustrating the general theory, recall the classical notion of topological entropy, and introduce its noncommutative counter part, Voiculescu entropy.

2.2.1 Endomorphisms of Cuntz algebras and a quantum shift

We will be interested in unital endomorphisms of Cuntz algebras (we will just call them endomorphisms of $\mathcal{O}_N$). Theorem 2.1.9 implies that they are automatically injective.

Theorem 2.2.1 ([Cu2]) There is a bijective correspondence between unitaries in $\mathcal{O}_N$ (that is, elements $U \in \mathcal{O}_N$ such that $U^*U = UU^* = 1_{\mathcal{O}_N}$) and endomorphisms of $\mathcal{O}_N$. It is given by the formulas

$$U_\rho = \sum_{i=1}^{N} \rho(S_i)S_i^*,$$

$$\rho_U(S_i) = US_i, \quad i = 1, \ldots, N$$

Proof Straightforward computation. Note that the fact that the second formula determines a *-homomorphism from $\mathcal{O}_N$ to $\mathcal{O}_N$ follows from the defining universal property of $\mathcal{O}_N$. □

A special class of endomorphisms of $\mathcal{O}_N$ is the class of so-called *permutation endomorphisms*, introduced in [Ka]. Let $k \in \mathbb{N}$ and let $\sigma : \mathcal{J}_k \to \mathcal{J}_k$ be a permutation. Define a unitary $U^\sigma \in \mathcal{O}_N$ by the formula

$$U^\sigma = \sum_{\mu \in \mathcal{J}_k} S_\mu S^*_{\sigma(\mu)}. \tag{2.2.1}$$

The endomorphism of $\mathcal{O}_N$ corresponding to U^σ via Theorem 2.2.1 will be denoted ρ_σ and called a permutation endomorphism.

Exercise 2.2.1 Check that the formula (2.2.1) indeed defines a unitary. Prove that each permutation endomorphism maps $\mathcal{C}_N$ into $\mathcal{C}_N$, and $\mathcal{F}_N$ into $\mathcal{F}_N$.

Exercise 2.2.2 Let U be a unitary in $\mathcal{F}_N \subset \mathcal{O}_N$. Prove that the associated endomorphism ρ_U maps $\mathcal{F}_N$ into $\mathcal{F}_N$ (note however that there exist endomorphisms ρ of $\mathcal{O}_N$ that map $\mathcal{F}_N$ into $\mathcal{F}_N$, but their associated unitaries U_ρ do not belong to $\mathcal{F}_N$ – see [CRS]).

Definition 2.2.2 The shift endomorphism (sometimes called also a quantum shift) is the endomorphism of $\mathcal{O}_N$ given by the formula:

$$\Phi(X) = \sum_{i=1}^{N} S_i X S_i^*, \quad X \in \mathcal{O}_N. \tag{2.2.2}$$

Exercise 2.2.3 Check that the formula (2.2.2) indeed defines an endomorphism of $\mathcal{O}_N$.

Proposition 2.2.3 The shift endomorphism is a permutation endomorphism given by the 'transposition' $\sigma : \mathcal{J}_2 \to \mathcal{J}_2, \sigma(i,j) = (j,i)$ for $i, j = 1, \ldots N$.

Proof It suffices to note that for each $i = 1, \ldots, N$

$$\rho_\sigma(S_i) = \sum_{j,k=1}^{N} S_j S_k (S_k S_j)^* S_i = \sum_{j=1}^{N} S_j S_i S_j^* = \Phi(S_i).$$

□

Theorem 2.2.4 Let T denote the left-sided shift on $\mathfrak{C}_N$ given by the formula

$$T((w_k)_{k=1}^{\infty})_l = w_{l+1}, \quad l \in \mathbb{N}.$$

The restriction of the shift endomorphism Φ to $\mathcal{C}_N$ is the map induced by T; precisely speaking (see Theorem 2.1.10 and the formula (2.1.1)):

$$\gamma \circ \Phi|_{\mathcal{C}_N} = \alpha_T \circ \gamma.$$

Proof This time it suffices to check that for each $\mu \in \bigcup_{k \in \mathbb{N}} \mathcal{J}_k$ the following equality holds:

$$\gamma(\Phi(S_\mu S_\mu^*)) = \alpha_T(Z_\mu),$$

where $Z_\mu \subset \mathfrak{C}_N$ is defined as in Theorem 2.1.10. We leave the details as an (enlightening!) exercise. □

Exercise 2.2.4 Is the identity map on $\mathcal{O}_N$ a permutation endomorphism? Is the shift Φ an automorphism (that is, is it surjective)? What other natural classes of unitaries in $\mathcal{O}_N$ one could consider in this context (for example using Exercise 2.1.4)?

In recent years Conti, Hong, and Szymański in a series of papers have undertaken a deep analysis of various classes of endomorphisms of $\mathcal{O}_N$ – for the information on this line of research we refer to the survey [CHS$_1$] and references therein.

2.2.2 The shift transformation on a graph C^*-algebra

Let $\Lambda = (V, E)$ be a finite directed graph with no sources (see Section 2.1.4). It is tempting and reasonable to study endomorphisms of $C^*(\Lambda)$ following the ideas of the previous section (see for example [CHS$_2$]); in general however, one cannot hope for a complete correspondence similar to that of Theorem 2.2.1. On the other hand, each $C^*(\Lambda)$ is equipped with a counterpart of the shift endomorphism of $\mathcal{O}_N$ that was introduced in Definition 2.2.2. Observe first that the space of infinite paths on Λ admits a natural continuous shift transformation, defined as follows:

$$T_\Lambda(f)(i) = f(i+1), \quad f \in \mathcal{P}_\Lambda^\infty, i \in \mathbb{N}_0.$$

Consider now $\Psi_\Lambda : C^*(\Lambda) \to C^*(\Lambda)$ defined by the formula:

$$\Psi_\Lambda(X) = \sum_{e \in E} S_e X S_e{}^*, \quad X \in C^*(\Lambda).$$

Proposition 2.2.5 The map Ψ_Λ introduced above is unital and completely positive; moreover, it leaves invariant the subalgebras $\mathcal{C}_\Lambda$ and $\mathcal{F}_\Lambda$. When restricted to $\mathcal{C}_\Lambda$, the map Ψ_Λ can be identified with the map induced by the shift on $\mathcal{P}_\Lambda^\infty$:

$$\gamma_\Lambda \circ \Psi_\Lambda = \alpha_{T_\Lambda} \circ \gamma_\Lambda.$$

Proof Unitality follows from the simple calculation:

$$\Psi_\Lambda(1) = \sum_{e \in E} S_e S_e^* = \sum_{v \in V} \sum_{e \in E : r(e) = v} S_e S_e{}^* = \sum_{v \in V} P_v = 1.$$

Complete positivity follows from the fact that each of the maps $X \mapsto S_e X S_e{}^*$ is completely positive. As it was for the endomorphism shift on Cuntz algebras, we leave the proof of the last part as an exercise. □

2.2.3 Classical topological entropy as defined by Rufus Bowen and extensions to topological pressure

Entropy is a numerical invariant associated with a dynamical system, very roughly speaking related to the system's chaotic/mixing properties. It was originally introduced by Kolmogorov (and Sinai) in 1950s for measurable dynamical systems, but later defined also for topological dynamical systems by Adler, Konheim, and McAndrew (see [NS] and references therein). Here we will recall an equivalent definition in the context of metric spaces, proposed by Bowen.

Let (X, d) be a compact metric space and let $T : X \to X$ be a continuous map. A finite set $F \subset\subset X$ is called (n, ϵ)-*spanning* for T ($n \in \mathbb{N}, \epsilon > 0$) if

$$\forall_{x \in X} \exists_{f \in F} \forall_{k=0,\ldots,n-1} d(T^k x, T^k f) < \epsilon.$$

(n, ϵ)-spanning sets always exist (Exercise! – it follows from compactness of X and continuity of T). Define

$$s(n, \epsilon) = \min\{\text{card}(F) : F - (n, \epsilon) - \text{ spanning subset for } T\}.$$

Definition 2.2.6 Topological entropy of T is the real number (possibly 0, possibly $+\infty$)

$$h_{\text{top}}(T) = \sup_{\epsilon > 0} \limsup_{n \to \infty} \frac{1}{n} \log s(n, \epsilon)$$

(in fact, one can replace in the definition the lim sup by lim inf, see Chapter 7 in [Wa]).

The idea behind Bowen's definition is related to approximating a given dynamical system (in the first n steps, up to ϵ) by a finite system. Below we will apply a similar scheme to define entropy for quantum dynamical systems. Before we do that let us first, however, compute the entropy of the one-sided shift on the Cantor set.

Proposition 2.2.7 Let T be the shift on $\mathfrak{C}_N$ introduced in Theorem 2.2.4. Then $h_{\text{top}}(T) = \log N$.

Proof Fix for the moment $k \in \mathbb{N}$ and put $\epsilon = \frac{1}{k}$. It is easy to check that $F_k \subset\subset \mathfrak{C}_N$ is an ϵ-net if and only if for any word w of length $k+1$ (built of letters from the alphabet $\{1, \ldots, N\}$) one can find a word in F_k such that its initial $(k+1)$-long segment coincides with w. Thus, the minimal cardinality of an ϵ-net is equal to N^{k+1}. A similar argument shows that for any $n \in \mathbb{N}$ the minimal cardinality of an (n, ϵ)-spanning set for T is equal to N^{k+1+n}. Thus we have

$$\limsup_{n\to\infty} \frac{1}{n} \log s(n,\epsilon) = \limsup_{n\to\infty} \frac{1}{n} \log N^{k+1+n}$$
$$= \limsup_{n\to\infty} \frac{n+k+1}{n} \log N = \log N$$

The fact that k was arbitrary implies that $h_{\text{top}}(T) = \log N$. □

The concept of topological entropy admits a natural extension to that of *topological pressure,* introduced by Ruelle and motivated by questions arising in statistical mechanics.

Definition 2.2.8 Let (X,d) be a compact metric space, let $T : X \to X$ be a continuous map and let $f \in C(X)$ be real-valued. The topological pressure of T at f is the number $P(T,f)$ defined as follows:

$$P(T,f) = \lim_{\epsilon\to 0} \limsup_{n\to\infty} \frac{1}{n} \log P_n(T,f,\epsilon),$$

where $P_n(T,f,\epsilon) = \inf\left\{\sum_{x\in F} \exp(-\sum_{i=0}^{n-1} f(T^i(x)))\right\}$, with the infimum taken over all F being (n,ϵ)-spanning subsets for T.

The expression $\sum_{x\in F} \exp(-\sum_{i=0}^{n-1} f(T^i(x)))$ is often called the *partition function*. More information about topological pressure can be found in Chapter 9 of [Wa] – note however that we are using here rather a convention of [NS] than this of [Wa] (hence, the minus sign before the sum in the exponent). It is easy to see that $P(T,0) = h_{\text{top}}(T)$, so that the topological pressure indeed generalizes the topological entropy; the two concepts share several properties.

Exercise 2.2.5 Consider the shift T on the Cantor set $\mathfrak{C}_N$, introduced in Theorem 2.2.4 and compute the pressure $p(T,1)$ and $p(T,\chi_{Z_1})$ (recall the notation for the cylinder sets introduced in Theorem 2.1.10).

2.2.4 Finite-dimensional approximations in the theory of C^*-algebras

Let A be a unital C^*-algebra. We will write $(\phi,\psi,M_n) \in CPA(\mathsf{A})$ to denote the facts that $\phi : M_n \to \mathsf{A}$, $\psi : \mathsf{A} \to M_n$ are unital completely positive (ucp) maps.

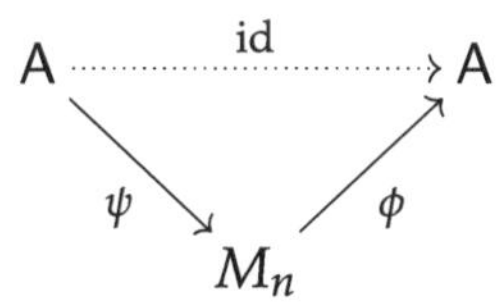

Further for $\Omega \subset\subset \mathsf{A}$ and $\epsilon > 0$ we will write $(\phi, \psi, M_n) \in CPA$ $(\mathsf{A}, \Omega, \epsilon)$ if $(\phi, \psi, M_n) \in CPA(\mathsf{A})$ and

$$\forall_{a\in\Omega} \, \|\phi \circ \psi(a) - a\| < \epsilon.$$

The last condition means that the above diagram is 'commutative on Ω up to ϵ'.

Definition 2.2.9 The algebra A is said to be *nuclear,* if $CPA(\mathsf{A}, \Omega, \epsilon) \neq \emptyset$ for any $\Omega \subset\subset \mathsf{A}, \epsilon > 0$. In such a case define

$$\mathrm{rcp}(\Omega, \epsilon) := \min\{n \in \mathbb{N} : \ (\phi, \psi, M_n) \in CPA(\mathsf{A}, \Omega, \varepsilon)\}.$$

Commutative C^*-algebras are nuclear – we will show this fact below. So are Cuntz algebras and more generally graph C^*-algebras (this can be shown in several ways – one approach is based on crossed-product-type constructions similar to those to be introduced in Section 2.4, another one on exploiting the properties of the algebra $\mathcal{F}_\Lambda$ and the fact that it is a fixed point algebra for a natural action of the circle $\mathbb{T}$ on $C^*(\Lambda)$). The reduced group C^*-algebra of a discrete group Γ is nuclear if and only if Γ is *amenable* (see Section 2.4). More information on C^*-approximations can be found in the excellent book [BrO]. We will later need the following simple observation.

Exercise 2.2.6 Let A, B be unital C^*-algebras, let $\gamma : \mathsf{A} \to \mathsf{B}$ be a ucp map and let $\Omega \subset\subset \mathsf{A}$. Show that if either A or B is nuclear, then for each $\epsilon > 0$ there exists $n \in \mathbb{N}$ and ucp maps $\psi : \mathsf{A} \to M_n$ and $\phi : M_n \to \mathsf{B}$ such that for all $a \in \Omega$ there is

$$\|\gamma(a) - (\phi \circ \psi)(a)\| < \epsilon.$$

2.2.5 Noncommutative topological entropy (Voiculescu topological entropy)

Definition 2.2.10 ([Vo]) Let (A, α) be a quantum dynamical system and assume that A is nuclear. Noncommutative topological entropy of α is the number

$$\operatorname{ht}\alpha = \sup_{\epsilon>0,\Omega\subset\subset \mathsf{A}} \limsup_{n\to\infty} \left(\frac{1}{n}\log \operatorname{rcp}(\Omega^{(n)},\epsilon)\right)$$

(here and below we use the notation $\Omega^{(n)} = \bigcup_{j=0}^{n-1} \alpha^j(\Omega)$).

Basic properties of $\operatorname{ht}\alpha$:

- ([Vo]) if $(\mathsf{A},\alpha) = (C(X),\alpha_T)$, and X is a compact metric space, then

$$\operatorname{ht}\alpha_T = h_{\text{top}}(T).$$

- ([Br]) ht does not increase when we pass to invariant subalgebras: if B is a C^*-subalgebra of A, such that $\alpha(\mathsf{B}) \subset \mathsf{B}$, then $\operatorname{ht}\alpha|_{\mathsf{B}} \leq \operatorname{ht}\alpha$.

None of the above facts is trivial; the second requires in fact a certain reformulation/extension of the definition (a subalgebra of a nuclear C^*-algebra need not be nuclear – can you see where the problem lies?). We will only prove the inequality

$$\operatorname{ht}\alpha_T \leq h_{\text{top}}(T). \tag{2.2.3}$$

Proof of the inequality (2.2.3) Assume that (X,d) is a compact metric space, $T : X \to X$ is a continuous map. Fix $\epsilon > 0$ and a finite set $\Omega \subset\subset C(X)$. Compactness of X and finiteness of Ω imply the existence of $\delta > 0$ such that

$$\forall_{f\in\Omega}\,\forall_{x,y\in X}\; d(x,y) < \delta \implies |f(x) - f(y)| < \epsilon.$$

Let $F = \{x_1,\ldots,x_k\}$ be an (n,δ)-spanning set for T such that $k = s(n,\delta)$. For each $i = 1,\ldots,k$ let

$$U_i = \{x \in X : \forall_{j=0,\ldots,n-1}\, d(T^j(x_i),T^j(x)) < \delta\}.$$

The family $(U_i)_{i=1}^k$ is an open cover of X. Let $(\varphi_i)_{i=1}^k$ be a corresponding partition of unity, that is, a family of functions in $C(X)$ such that each φ_i is non-negative, supported in U_i and moreover $\sum_{i=1}^k \varphi_i = 1$.

Define linear maps $\psi : C(X) \to M_k$ and $\phi : M_k \to C(X)$ by the formulas

$$\psi(f) = \operatorname{diag}[f(x_1),\ldots,f(x_k)], \quad f \in C(X),$$

$$\phi([a_{ij}]_{i,j=1}^k) = \sum_{i=1}^k a_{ii}\varphi_i, \quad [a_{ij}]_{i,j=1}^k \in M_k.$$

It is easy to see that the maps ψ and ϕ are linear, unital, and positive. Remark 2.1.6 implies that $(\phi, \psi, M_k) \in CPA(C(X))$.

Let then $g \in \Omega^{(n)} = \bigcup_{j=0}^{n-1} \alpha_T^j(\Omega)$, say $g = \alpha_T^j(f)$ for certain $j \in \{0, \dots, n-1\}$, $f \in \Omega$. Then for any $x \in X$

$$\begin{aligned}|(\phi \circ \psi)(g)(x) - g(x)| &= |\sum_{i=1}^k \varphi_i(x) g(x_i) - g(x)| \\ &= |\sum_{i=1}^k \varphi_i(x)(g(x_i) - g(x))| \leq \sum_{i=1}^k \varphi_i(x)|g(x_i) - g(x)| \\ &= \sum_{i=1,\dots,k \text{ s.t.} x \in U_i} \varphi_i(x)|g(x_i) - g(x)| < \epsilon,\end{aligned}$$

where the last inequality follows from the fact that $g(x_i) - g(x) = f(T^j(x_i)) - f(T^j(x))$ and if $x \in U_i$, then $d(T^j(x_i), T^j(x)) < \delta$. Thus, we have shown that $(\phi, \psi, M_k) \in CPA(C(X), \Omega^{(n)}, \epsilon)$, so that $\text{rcp}(\Omega^{(n)}, \epsilon) \leq k = s(n, \delta)$.

The inequality (2.2.3) easily follows. □

Exercise 2.2.7 Show that if X is compact, then $C(X)$ is nuclear. Do we need to assume that X is metrisable?

Exercise 2.2.8 Let (A, α) be a noncommutative dynamical system with A nuclear and let $n \in \mathbb{N}$. Prove that $M_n(\mathsf{A})$ is nuclear and that $\text{ht}\, \alpha^{(n)} = \text{ht}\, \alpha$.

Providing an elementary proof of the inequality

$$\text{ht}\, \alpha_T \geq h_{\text{top}}(T)$$

turns out to be surprisingly difficult – still the only available argument is the original one published in [Vo]. It uses a so-called *CNT entropy* (due to Connes–Narnhofer–Thirring), a noncommutative version of the classical measure entropy. It is also not clear whether one can replace in Voiculescu's definition the lim sup by lim inf.

In the following we will need one more important property of the Voiculescu entropy. It allows us not to check all finite subsets of a given C^*-algebra when computing the entropy.

Proposition 2.2.11 (Kolmogorov–Sinai property, [Vo]) Let A be a nuclear C^*-algebra, $\alpha : \mathsf{A} \to \mathsf{A}$ a unital *-homomorphism. If $(\Omega_i)_{i\in I}$ is a family of finite subsets of A such that $\bigcup_{i\in I, n\in\mathbb{N}} \Omega_i^{(n)}$ is linearly dense in A, then

$$\mathrm{ht}\,\alpha = \sup_{\epsilon>0, i\in I} \limsup_{n\to\infty} \left(\frac{1}{n}\log \mathrm{rcp}(\Omega_i^{(n)}, \epsilon)\right).$$

Note that in particular we can apply the above result in the situation where $\bigcup_{i\in I}\Omega_i$ is linearly dense in A.

Finally, we introduce the notion of noncommutative topological pressure owing to Neshveyev and Størmer. Let A be a nuclear C^*-algebra, $\alpha : \mathsf{A} \to \mathsf{A}$ a unital *-homomorphism and $a \in \mathsf{A}$ be self-adjoint. Define for each $n \in \mathbb{N}$ the element $a^{(n)} = \sum_{i=0}^{n-1} \alpha^i(a)$ and the *noncommutative partition function* ($\epsilon > 0$, $n \in \mathbb{N}$)

$$Z_{\alpha,n}(a, \Omega, \epsilon) = \inf\{\mathrm{Tr}\, \mathrm{e}^{\psi(a^{(n)})} : (\phi, \psi, M_k) \in CPA(\mathsf{A}, \Omega^{(n)}, \epsilon)\},$$

where Tr denotes the canonical (not normalized) trace on M_k (that is, $\mathrm{Tr}(q) = 1$ for any minimal projection $q \in M_k$), and for a matrix $M \in M_k$ we write $\mathrm{e}^M = \sum_{n=0}^{\infty} \frac{M^n}{n!}$. Define

$$P_\alpha(a, \Omega, \varepsilon) = \limsup_{n\to\infty} \frac{1}{n} \log\left(Z_{\alpha,n}(a, \Omega, \varepsilon)\right),$$

and the *noncommutative pressure*

$$P_\alpha(a) = \sup_{\varepsilon>0, \Omega\subset\subset \mathsf{A}} P_\alpha(a, \Omega, \varepsilon).$$

Again, as in the classical case, for any quantum dynamical system (A, α) we have

$$\mathrm{ht}\,\alpha = P_\alpha(0).$$

Further, for any classical dynamical system (X, T) and a real-valued $f \in C(X)$ we have $P(T, f) = P_{\alpha_T}(f)$; the noncommutative dynamical pressure satisfies also a version of the Kolmogorov–Sinai property.

Finally, note that both the notions of noncommutative topological entropy and pressure can be naturally extended to unital completely positive maps of nuclear C^*-algebras (which we will use in the following section). More generally, one can study similar ideas for other mathematical categories, as soon as one defines a natural notion of a 'finite-dimensional approximation up to ϵ' and a method of defining the size of this approximation. For an application of such an idea to the study of Banach space contractions we refer to [KL].

Finally, as we already mentioned that proving that if T is a continuous transformation of a compact space X then the Voiculescu entropy $\mathrm{ht}\,\alpha_T$ is equal to $h_{\mathrm{top}}(T)$ is difficult, it is a natural place to state a related open problem.

Problem 2.2.12 Let X be a compact space and let $\phi : C(X) \to C(X)$ be a unital, positive map (recall it is automatically completely positive). Then Downarowicz and Frej propose in [DF] three natural definitions of topological entropy of ϕ, very roughly speaking in terms of open covers or separating subsets involving both X and finite families of functions in $C(X)$, and show that they lead to the same numerical value, which we denote by $h_{DF}(\phi)$. It is easy to see, essentially using the methods presented above, that $\mathrm{ht}\,\phi \leq h_{DF}(\phi)$. Do we always have $\mathrm{ht}\,\phi = h_{DF}(\phi)$?

2.3 Voiculescu Entropy of the Shift and other Examples

The main aim of this section is to compute explicitly the Voiculescu entropy and noncommutative topological pressure in some examples. We will on one hand present cases where the method of looking at commutative subalgebras yields the final answer and on the other explain why this technique may sometimes fail.

2.3.1 Estimating the Voiculescu entropy of the permutative endomorphisms of Cuntz algebras

We begin by establishing the following estimate for the Voiculescu entropy of the permutative endomorphisms of $\mathcal{O}_N$.

Theorem 2.3.1 ([SZ$_2$]) Fix $N \in \mathbb{N}$, let $k \in \mathbb{N}$ and let $\sigma : \mathcal{J}_k \to \mathcal{J}_k$ be a permutation. Denote the permutative endomorphism of $\mathcal{O}_N$ corresponding to σ (via the formula (2.2.1) and Theorem 2.2.1) by ρ. Then

$$\mathrm{ht}\,\rho \leq (k-1)\log N. \tag{2.3.1}$$

The proof will be divided into a series of lemmas, in which N and k will be fixed (and ρ will be as above). For $i, j \in \mathbb{N}$ put

$$A_{i,j} = \{S_\mu S_\nu^* : \mu \in \mathcal{J}_i, \nu \in \mathcal{J}_j\}, \quad F_{i,j} = \operatorname{Lin} A_{i,j}.$$

Lemma 2.3.2 For any $i, j, m \in \mathbb{N}$

$$\rho^m(A_{i,j}) \subset F_{i+m(k-1),j+m(k-1)}.$$

Proof It suffices to consider the case of $m = 1$ (do you see why?). Let then $\mu \in \mathcal{J}_i$, $\nu \in \mathcal{J}_j$ and note that

$$\rho(S_\mu) = U_\sigma S_{\mu_1} \cdots U_\sigma S_{\mu_i} \in F_{k,k-1} \cdots F_{k,k-1} \subset F_{k+i-1,k-1}.$$

Similarly

$$\rho(S_\nu)^* \in F_{k-1,k+j-1}.$$

Hence

$$\rho(S_\mu S_\nu^*) \in F_{k+i-1,k+j-1}.$$

□

For each $l \in \mathbb{N}$ introduce the following (injective) *-homomorphism $\Psi_l : \mathcal{O}_N \to M_{N^l}(\mathcal{O}_N)$:

$$\Psi_l(X) = \sum_{\mu,\nu \in \mathcal{J}_l} e_{\mu,\nu} \otimes S_\mu^* X S_\nu, \quad X \in \mathcal{O}_N. \tag{2.3.2}$$

We identify above the algebra $M_{N^l}(\mathcal{O}_N)$ with the tensor product $M_{N^l} \otimes \mathcal{O}_N$, index the set $\{1, \ldots, N^l\}$ by indices in $\mathcal{J}_l$ and denote by $e_{\mu,\nu}$ the relevant matrix unit – the matrix whose only non-zero entry is 1 in the 'μ'-th row and 'ν'-th column.

Exercise 2.3.1 Check that the formula (2.3.2) indeed defines a unital *-homomorphism.

Lemma 2.3.3 Let $l, i, j \in \mathbb{N}$, $l \geq i > j$. Let $X \in F_{i,j}$. Then for each $\mu \in \mathcal{J}_{i-j}$ there exists $T_\mu \in M_{N^l}$ such that $\|T_\mu\| \leq \|X\|$ and

$$\Psi_l(X) = \sum_{\mu \in \mathcal{J}_{i-j}} T_\mu \otimes S_\mu. \tag{2.3.3}$$

Proof To prove the existence of T_μ (without the norm condition), it suffices to see what happens for $X \in A_{i,j}$. For each $\nu \in \mathcal{J}_i$, $\kappa \in \mathcal{J}_j$ we have

$$\Psi_l(S_\nu S_\kappa^*) = \sum_{\beta,\delta\in\mathcal{J}_l} e_{\beta,\delta} \otimes S_\beta^* S_\nu S_\kappa^* S_\delta = \sum_{\beta'\in\mathcal{J}_{l-i},\delta'\in\mathcal{J}_{l-j}} e_{\nu\beta',\kappa\delta'} \otimes S_{\beta'}^* S_{\delta'}$$
$$= \sum_{\mu\in\mathcal{J}_{i-j}} \Big(\sum_{\beta'\in\mathcal{J}_{l-i}} e_{\nu\beta',\kappa\beta'\mu}\Big) \otimes S_\mu.$$

It remains to show the norm estimate for each matrix T_μ appearing in the formula (2.3.3). We have:

$$\| \sum_{\mu\in\mathcal{J}_{i-j}} T_\mu \otimes S_\mu\|^2 = \| \sum_{\mu\in\mathcal{J}_{i-j}} (T_\mu \otimes S_\mu)^* \sum_{\nu\in\mathcal{J}_{i-j}} (T_\nu \otimes S_\nu)\|$$
$$= \| \sum_{\mu,\nu\in\mathcal{J}_{i-j}} T_\mu^* T_\nu \otimes S_\mu^* S_\nu\| = \| \sum_{\mu\in\mathcal{J}_{i-j}} T_\mu^* T_\mu\|,$$

so for each $\mu \in \mathcal{J}_{i-j}$

$$\|T_\mu\|^2 = \|T_\mu^* T_\mu\| \le \| \sum_{\nu\in\mathcal{J}_{i-j}} T_\nu \otimes S_\nu\|^2 = \|\Psi_l(X)\|^2. \tag{2.3.4}$$

As Ψ_l is an injective *-homomorphism, we obtain

$$\|\Psi_l(X)\| = \|X\|,$$

which together with the estimate (2.3.4) ends the proof. □

Analogous lemmas can be established for the case $i = j$ and $i < j$ (always keeping $i, j \le l$) – we suggest formulating and proving them as an exercise.

We are ready for the proof of Theorem 2.3.1.

The proof of Theorem 2.3.1 For each $l \in \mathbb{N}$ let

$$\Omega_l = \bigcup_{p,q=1}^{l} A_{p,q}.$$

Fix $l \in \mathbb{N}$, $\epsilon > 0$. As $\mathcal{O}_N$ is nuclear, there exists a triple $(\phi_0, \psi_0, M_{C_l})$ $\in CPA(\mathcal{O}_N, \Omega_l, \frac{1}{4N^l}\epsilon)$. Fix now $n \in \mathbb{N}$ and put

$$\Omega_l^{(n)} = \bigcup_{j=0}^{n-1} \rho^j(\Omega_l).$$

Let $m = n(k-1)+l$. Apply now the conclusion of Exercise 2.2.6 to the map $\Psi_m^{-1} : \Psi_m(\mathcal{O}_N) \to \mathcal{O}_N$. There exist $d \in \mathbb{N}$ and unital completely positive maps $\gamma : \Psi_m(\mathcal{O}_N) \to M_d$ and $\eta : M_d \to \mathcal{O}_N$ such that for all $a \in \Psi_m(\Omega_l^{(n)})$ we have

$$\|\eta \circ \gamma(a) - \Psi_m^{-1}(a)\| < \frac{\epsilon}{2}.$$

Let $\widetilde{\gamma} : M_{N^m} \otimes \mathcal{O}_N \to M_d$ be a unital completely positive extension of γ (see Theorem 2.1.7). Consider the following diagram:

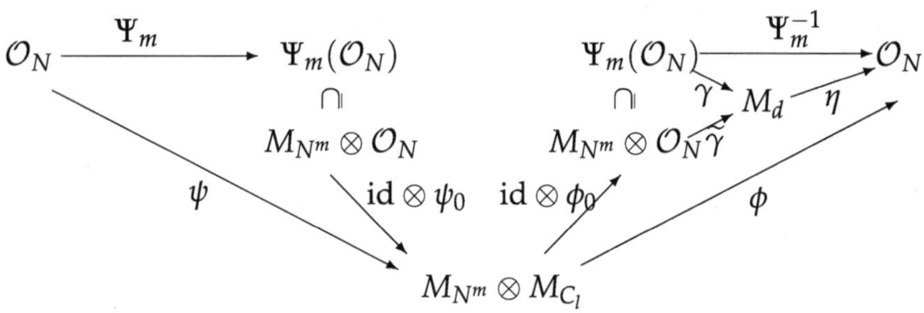

Let $X \in A_{p,q}$ $(p, q \leq l)$ and let $j \in \mathbb{N}_0$, $j \leq n-1$. Then

$$\begin{aligned}
&\|\phi \circ \psi(\rho^j(X)) - \rho^j(X)\| \\
&\quad = \|\eta \circ \widetilde{\gamma} \circ (\mathrm{id} \otimes \phi_0 \circ \psi_0) \circ \Psi_m(\rho^j(X)) - (\Psi_m^{-1} \circ \Psi_m)(\rho^j(X))\| \\
&\quad = \|\eta \circ \widetilde{\gamma} \circ (\mathrm{id} \otimes \phi_0 \circ \psi_0) \circ \Psi_m(\rho^j(X)) - \eta \circ \widetilde{\gamma} \circ \Psi_m(\rho^j(X)) \\
&\qquad\qquad + \eta \circ \widetilde{\gamma} \circ \Psi_m(\rho^j(X)) - (\Psi_m^{-1} \circ \Psi_m)(\rho^j(X))\| \\
&\qquad \leq \|(\mathrm{id} \otimes \phi_0 \circ \psi_0) \circ \Psi_m(\rho^j(X)) - \Psi_m(\rho^j(X))\| + \frac{\epsilon}{2}.
\end{aligned}$$

Lemma 2.3.2 implies that $\rho^j(X) \in F_{p+j(k-1),q+j(k-1)}$. Assume for example that $p > q$. Then, as $\|X\| \leq 1$, it follows from Lemma 2.3.3 that

$$\begin{aligned}
\|(\phi \circ \psi)(\rho^j(X)) - \rho^j(X)\| &\leq \Big\| \sum_{\mu \in \mathcal{J}_{p-q}} T_\mu \otimes ((\phi_0 \circ \psi_0)(S_\mu) - S_\mu) \Big\| + \frac{\epsilon}{2} \\
&\leq \sum_{\mu \in \mathcal{J}_{p-q}} \|T_\mu\| \, \|(\phi_0 \circ \psi_0)(S_\mu) - S_\mu\| < N^{p-q} \frac{\epsilon}{4N^l} + \frac{\epsilon}{2} = \epsilon
\end{aligned}$$

(as $S_\mu \in A_{p-q,0} \in \Omega_l$). Hence, we have shown that

$$(\phi, \psi, M_{N^m} \otimes M_{C_l}) \in CPA(\mathcal{O}_N, \Omega_l^{(n)}, \epsilon). \tag{2.3.5}$$

This implies that $\mathrm{rcp}(\Omega_l^{(n)}, \epsilon) \leq C_l N^m$,

$$\log \mathrm{rcp}(\Omega_l^{(n)}, \epsilon) \leq C_l + m \log N = C_l + ((k-1)n + l) \log N$$

and further

$$\limsup_{n\to\infty} \left(\frac{1}{n} \log \mathrm{rcp}(\Omega_l^{(n)}, \epsilon) \right) \leq (k-1) \log N.$$

Kolmogorov–Sinai property of the Voiculescu entropy (Proposition 2.2.11) ends the proof. □

The above result is a simplified version of Theorem 2.2 from [SZ$_2$]. The proof is partially inspired by the article [BG].

2.3.2 The Voiculescu entropy of the noncommutative shift and related questions

Theorem 2.3.1 allows us in particular to obtain the following result, originally proved, via different methods, in [Ch].

Theorem 2.3.4 ([Ch]) Let $N \in \mathbb{N}, N \geq 2$ and let $\Phi : \mathcal{O}_N \to \mathcal{O}_N$ be the shift endomorphism. Then $\mathrm{ht}\, \Phi = \log N$.

Proof Theorem 2.3.1 and Propositions 2.2.3 imply that

$$\mathrm{ht}\, \Phi \leq \log N.$$

On the other hand Theorem 2.2.4 and the monotonicity of the Voiculescu entropy allow us to deduce that

$$\mathrm{ht}\, \Phi \geq h_{\mathrm{top}}(T),$$

where T denotes the shift on $\mathfrak{C}_N$. As $h_{\mathrm{top}}(T) = \log N$ (Proposition 2.2.7), the proof is finished. □

The proof of the above theorem is a standard example of computing the Voiculescu entropy via 'classical subsystems'. Even for relatively simple permutative endomorphisms of Cuntz algebras this method may need to be fine-tuned, as the following exercise shows (we will later see in Theorem 2.3.7 that in other cases this approach may simply fail).

Exercise 2.3.2 Let $N = 2$. Consider the endomorphism $\rho : \mathcal{O}_2 \to \mathcal{O}_2$ given by the prescription

$$\rho(S_1) = S_1 S_2 S_1^* + S_1 S_1 S_2^*, \quad \rho(S_2) = S_2.$$

(i) Show that ρ is a permutative endomorphism. Deduce further that $\mathrm{ht}\,\rho \leq \log 2$.

(ii) Find a continuous map T on the Cantor set $\mathfrak{C}_2$ such that $\rho|_{\mathcal{C}_2}$ corresponds to α_T (in the sense used in Theorem 2.2.4).

(iii) Show that $h_{\mathrm{top}}(T) = 0$.

(iv) Try to think how one could show that

$$\mathrm{ht}\,\rho = \log 2. \tag{2.3.6}$$

Equality (2.3.6) was established in [SZ$_2$].

In view of the discussion above and of Exercise 2.2.2 it is natural to raise the following open problem.

Problem 2.3.5 Does there exist a unitary $U \in \mathcal{F}_N \subset \mathcal{O}_N$ such that $\mathrm{ht}\,\rho_U > \mathrm{ht}\,\rho_U|_{\mathcal{F}_N}$? The evidence so far shows that the answer is likely negative.

2.3.3 Generalizations to graph C^*-algebras and to noncommutative topological pressure

Theorem 2.3.4 has a far-reaching generalization, established in [SZ$_1$], the special case of which we formulate here in the framework of C^*-algebras associated to finite graphs. Recall that if Λ is a finite graph with no sources then $\mathcal{C}_\Lambda$ denotes the commutative C^*-subalgebra of $C^*(\Lambda)$ isomorphic (via the map γ_Λ) to the algebra of continuous functions $C(\mathcal{P}_\Lambda^\infty)$ on the infinite path space of Λ; the isomorphism may be used to identify $\Psi_\Lambda|_{\mathcal{C}_\Lambda}$ with the *-homomorphism induced by the shift map T_Λ on $\mathcal{P}_\Lambda^\infty$ (see Proposition 2.2.5).

Theorem 2.3.6 Let Λ be a row-finite graph with no sources and let $f \in C(\mathcal{P}_\Lambda^\infty)$ be real-valued. Then the following equality holds:

$$P_{\Phi_\Lambda}(\gamma_\Lambda(f)) = P(T_\Lambda, f).$$

The key arguments needed in the proof of the above theorem are contained in the proof of Theorem 2.3.1; the non-trivial part is showing the inequality '$\leq$', as the other one follows from monotonicity of the noncommutative topological pressure. We recommend the readers a careful analysis of two special cases: the case where Λ is the graph with one vertex and N edges, that is, $C^*(\Lambda) = \mathcal{O}_N$, and the case where $f = 0$ (so that we simply ask about the Voiculescu entropy).

Exercise 2.3.3 Use the above theorem and information contained in previous sections to reproduce the main result of [BG]. Specifically, let $k \in \mathbb{N}$ and $M \in M_k(\{0,1\})$ be a matrix such that the associated finite graph Λ_M has no sources. Then the Voiculescu entropy of the 'generalized shift' Φ_{Λ_M} on the Cuntz–Krieger algebra $C^*(\Lambda_M)$ is equal to the spectral radius of the matrix M, that is, the limit $\lim_{n\to\infty} \|M^n\|^{\frac{1}{n}}$.

2.3.4 Automorphism whose Voiculescu entropy is genuinely noncommutative

Theorem 2.3.4 and Exercise 2.3.2 may seem to suggest that the Voiculescu entropy could be always simply equal to the supremum of classical entropies of *all* of its classical subsystems.

Define for a quantum dynamical system (A, α) the following number:

$$\mathrm{ht}_c(\alpha) = \sup \Big\{ \mathrm{ht}(\alpha|_{\mathsf{C}}) : \mathsf{C}\text{—commutative } \alpha\text{-invariant } C^*\text{-subalgebra of } \mathsf{A} \Big\}.$$

General properties of the Voiculescu entropy imply that we always have $\mathrm{ht}\,\alpha \geq \mathrm{ht}_c(\alpha)$. An example of an automorphism of a nuclear C^*-algebra for which the converse inequality does not hold was given in [Sk$_1$]. We will now briefly describe the relevant model and list the properties that are crucial for the required argument.

Fix a set $X \subset \mathbb{N}_0$. Consider $\mathsf{A}(X)$, the universal C^*-algebra of the *bit-stream* associated to X: $\mathsf{A}(X)$ is generated by the family of operators $\{s_i : i \in \mathbb{Z}\}$, which are self-adjoint unitaries (that is, $s_i^2 = I$, $s_i = s_i^*$ for each $i \in \mathbb{Z}$ – such operators are often called symmetries) and satisfy the following commutation relations:

$$s_i s_j = (-1)^{\chi_X(|i-j|)} s_j s_i, \quad i,j \in \mathbb{Z},$$

where $\chi_X : \mathbb{N}_0 \to \{0,1\}$ is the characteristic function of X. Note that these relations simply mean that the symmetries in question either commute, or anti-commute, depending on their relative position with respect to X. As it is only the relative position (that is, the distance from i to j) which matters, there exists an automorphism σ_X of $\mathsf{A}(X)$, called the *binary shift*, determined by the prescription

$$\sigma_X(s_i) = s_{i+1}, \quad i \in \mathbb{Z}.$$

The algebra $\mathsf{A}(X)$ was first introduced and studied by Powers. It can be identified with a twisted version of the C*-algebra of the compact group $\bigoplus_{\mathbb{Z}} \mathbb{Z}_2$. In particular it is nuclear, and possesses a faithful tracial state τ, uniquely determined by the formula

$$\tau(s_i) = \begin{cases} 1 & \text{if} \quad i = 0, \\ 0 & \text{if} \quad i \in \mathbb{Z} \setminus \{0\}. \end{cases}$$

In general the properties of the quantum dynamical system $(\mathsf{A}(X), \sigma_X)$ depend on the choice of X, but one can show that for almost every (with respect to the Bernoulli measure on $2^{\mathbb{N}} \approx \prod_{n \in \mathbb{N}} \{0,1\}$) nonperiodic $X \subset \mathbb{N}$ the following properties hold:

(i) τ is the unique σ_X-invariant state on $\mathsf{A}(X)$;
(ii) the CNT entropy of σ_X with respect to the trace τ vanishes: $h_\tau(\sigma_X) = 0$;
(iii) the CNT entropy of $\sigma_X \otimes \sigma_X$ with respect to the trace $\tau \otimes \tau$ does not vanish: $h_{\tau\otimes\tau}(\sigma_X \otimes \sigma_X) = \log 2$.

Note that in particular the binary shifts provide counterexamples for the additivity of the CNT entropy: $h_{\tau\otimes\tau}(\sigma_X \otimes \sigma_X) \neq h_\tau(\sigma_X) + h_\tau(\sigma_X)$. For the proof of these properties we refer to Chapter 12 of [NS].

Using these properties and some careful, but not too difficult von Neumann algebra type arguments (see Section 2.5) one can prove the following result.

Theorem 2.3.7 ([Sk$_1$]) For almost every nonperiodic set $X \subset \mathbb{N}$ the binary shift σ_X satisfies $\mathrm{ht}_c(\sigma_X) = 0 < \frac{\log 2}{2} \leq \mathrm{ht}(\sigma_X)$.

Sketch The upper estimate follows immediately from property (iii) above, subadditivity of Voiculescu's topological entropy with respect to tensor products and the fact that it dominates the CNT entropy.

For the lower estimate we assume that C be a commutative σ-invariant C*-subalgebra of A, write $\sigma_X|_{\mathsf{C}} = \sigma_{\mathsf{C}}$ and note that due to the classical variational principle (which says that the topological entropy for a classical dynamical system is equal to the supremum of the corresponding measure entropies with respect to invariant measures for the system, see Chapter 8 of [Wa]) it suffices to show that the classical measure entropy $h_\phi(\sigma_{\mathsf{C}}) = 0$ for every σ_{C}-invariant state ϕ on C (which is in fact a measure on the spectrum of C). Using a straightforward ergodic type argument using property (i) above, we see that ϕ must be equal to $\tau|_{\mathsf{C}}$. This would have ended the proof if we knew that the CNT entropy is monotone under passing to invariant subalgebras (see property (ii) above), but this holds true only for so-called expected subalgebras. Thus, we need to pass to the von Neumann algebra set-up and exploit the fact that τ is tracial, implying that suitable conditional expectations (that is, norm-one surjective projections on the von Neumann subalgebras in question, preserving the trace τ) always exist. □

Note that the analogous question for the CNT entropy of the automorphisms of the hyperfinite II_1-factor R remains open – in other words we do not know if there exists an automorphism of R whose CNT entropy (with respect to the canonical trace τ on R) is greater than the supremum of respective measure entropies over the restrictions to all invariant commutative von Neumann subalgebras of R.

Theorem 2.3.7 implies also a necessity of looking for other techniques of estimating the Voiculescu entropy from below. Possible ways are related to investigating the classical entropy of the induced dynamics of the state space of A ([Ke]) or exploiting the connections to the index theory of von Neumann algebras ([Sk_2], see also [Hi]).

2.3.5 Automorphism that leaves no non-trivial abelian subalgebras invariant

In view of the discussion above it might be natural to ask whether for any noncommutative dynamical system (A, α) there exists a non-trivial (that is, different from $\mathbb{C}1_{\mathsf{A}}$) commutative C^*-subalgebra $\mathsf{C} \subset \mathsf{A}$ left globally invariant by α. The following example, inspired by a suggestion of Kümmerer, shows that this is not the case.

Before we formulate the example we need to recall some facts regarding the spectrum of unitary operators (for the details see the books [He] and [Na]). Every unitary operator U on a separable Hilbert space is isomorphic to a countable direct sum of the operators of the form $M_z \in L^2(\mathbb{T}, \mu)$, where μ is a measure on the circle, and $(M_z(f))(t) = tf(t)$ for any $t \in \mathbb{T}, f \in L^2(\mathbb{T}, \mu)$. As each such measure μ decomposes into a discrete part, a part which is absolutely continuous with respect to the Lebesgue measure and the remainder (the singular part), we obtain the analogous decomposition of U; moreover, this decomposition is unique in a natural sense. The discrete part corresponds to the *discrete spectrum* of U (that is, the collection of eigenvalues of U). A standard example of a unitary with the Lebesgue continuous spectrum is the two-sided shift on $\ell^2(\mathbb{Z})$, which is isomorphic to M_z acting on $L^2(\mathbb{T}, \lambda)$ with λ the Lebesgue measure. Finally, we say that U has *purely singular spectrum* if all the measures μ appearing in the decomposition are singular. The uniqueness mentioned above implies that such a U admits no eigenvalues and no invariant subspace on which it has Lebesgue continuous spectrum (examples can be constructed of course by looking at M_z on $L^2(\mathbb{T}, \mu)$ for a fixed singular measure on $\mathbb{T}$).

Theorem 2.3.8 Let H be a separable infinite-dimensional Hilbert space and $U \in B(\mathsf{H})$ be a unitary with purely singular continuous spectrum. Consider the C^*-algebra $\mathsf{A} = \mathcal{K}_\mathsf{H} + \mathbb{C}1 \subset B(\mathsf{H})$ and the automorphism of A given by $\alpha(x) = U^*xU$, $x \in \mathsf{A}$. Then $\mathbb{C}1$ is the only commutative non-trivial C^*-subalgebra left invariant by α.

Proof Let U be as above. Suppose that C is a commutative C^*-subal–gebra of A left invariant by α. Then $\mathsf{C} \cap \mathcal{K}_\mathsf{H}$ is also left invariant by α; therefore, we may assume that $\mathsf{C} \subset \mathcal{K}_\mathsf{H}$ (thus, we need to deduce that $\mathsf{C} = \{0\}$).

We begin by understanding how commutative subalgebras C of $\mathcal{K}_\mathsf{H}$ look like. Any operator $x \in \mathsf{C}$ is normal. As it is also compact, it is of the form $\sum_{i=1}^k \lambda_i P_i$, where P_i are finite-rank projections and λ_i are non-zero complex numbers, and either $k = \infty$, in which case the sequence $(\lambda_i)_{i=0}^\infty$ converges to 0, or $k \in \mathbb{N}$ (we can assume that $x \neq 0$). Functional calculus implies that each P_i belongs to C. This means that C is spanned (in the sense described above) by its projections. In the next step we consider the family

$$\mathcal{P}_{min}^{\mathsf{C}} := \{p \in \mathsf{C} : p \text{ minimal non-zero projection in } \mathsf{C}\}.$$

Elements of $\mathcal{P}_{min}^{\mathsf{C}}$ are mutually orthogonal (can you see why?) and can be listed in the sequence $(Q_i)_{i=1}^\infty$ (remember that H is separable)[1]. It follows that $\mathsf{C} = \{\sum_{i=1}^\infty \lambda_i Q_i : (\lambda_i)_{i=0}^\infty \in c_0\}$.

[1]Strictly speaking the sequence could be finite, but then the following arguments would also work – it would suffice to consider the case of a finite cycle.

We will now use the assumption that α leaves C invariant. This implies that for each $i \in \mathbb{N}$ there exists $k_i \in \mathbb{N}$ such that $\alpha(Q_i) = Q_{k_i}$. Indeed, $\alpha(Q_i)$ must be a projection in C (of the same rank as Q_i) and considering α^{-1} we see that $\alpha(Q_i)$ is also minimal (in C). The map $\mathbb{N} \ni i \mapsto k_i$ is a permutation of $\mathbb{N}$, which we will denote by σ.

Recall now the concrete form of α. The equality $\alpha(Q_i) = Q_{k_i}$ means that the unitary U^* (whose spectral type is the same as that of U) maps the space $Q_i\mathsf{H}$ onto $Q_{\sigma(i)}\mathsf{H}$. We then need to consider two cases. First suppose that σ contains a finite cycle, say $l_1, \ldots, l_k$. Then the space $\mathsf{K} := \bigoplus_{i=1}^k Q_{l_i}\mathsf{H}$ is a finite-dimensional subspace of H invariant for U and the spectral theorem for matrices implies that U must have an eigenvector inside K. This is however impossible, as the point spectrum of U^* is empty.

It remains then to consider the case when σ contains an infinite orbit. Choose k lying in such an orbit and a norm-one vector in $Q_k\mathsf{H}$, say ξ. Put $\xi_k = U^k\xi$ for all $\xi \in \mathbb{Z}$. Then the space $\mathsf{K}' := \overline{\mathrm{Lin}}\{\xi_k : k \in \mathbb{Z}\}$ is left invariant by U, and as $U\xi_k = \xi_{k+1}$ ($k \in \mathbb{Z}$), the restriction of U to K' is unitarily equivalent to the usual two-sided shift S on $\ell^2(\mathbb{Z})$. This contradicts the remarks before the theorem. □

It may seem that the last theorem is likely to provide an alternative to Theorem 2.3.7 in offering an example of an automorphism of a nuclear C^*-algebra whose Voiculescu entropy is strictly greater than the supremum of classical entropies of all of its classical subsystems. This is however not true, as any automorphism α of $\mathcal{K}_\mathsf{H} + \mathbb{C}1$ has Voiculescu entropy equal to 0. This is proved in Example 6.2.10 (i) of [NS] – the argument is based on the fact that it suffices to consider the sets Ω supported on finite-dimensional subspaces of H, say $p\mathsf{H}$, where p is a finite rank projection. Then for $n \in \mathbb{N}$ the set $\bigcup_{k=0}^{n-1} \alpha^k(\Omega)$ is supported in the space $p_n\mathsf{H}$, where $p_n = \bigvee_{k=0}^{n-1} \alpha^k(p)$ and one can build an (exact) approximation via $p_nB(\mathsf{H})p_n$ (strictly speaking we also need to cater for $\mathbb{C}1$, but that is easy). The latter algebra has rank equal to the rank of p_n, which cannot be greater than n times rank of p. Thus, the size of approximations grows at most linearly and $\mathrm{ht}\,\alpha = 0$.

2.4 Crossed Products and the Entropy

In this section we introduce the crossed product construction, which in its simplest form we will encounter here may be viewed as means of 'making' a given automorphism of a C^*-algebra inner – albeit in a larger C^*-algebra. Then we discuss a natural extension procedure for a map on a C^*-algebra commuting with the initial automorphism to the crossed product and show that this procedure does not change the Voiculescu entropy.

2.4.1 Crossed products

Let (α, A) be a noncommutative dynamical system, with $\alpha \in \mathrm{Aut}(\mathsf{A})$. Recall that an automorphism α is called *inner* if there exists a unitary $u \in \mathsf{A}$ such that

$$\alpha(a) = u^* a u, \quad u \in \mathsf{A}.$$

We then write $\alpha = \mathrm{Ad}\, u$. Note that usually not all automorphisms of a given A are inner (consider the example of $\mathsf{A} = C(X)$ for a compact space X).

Below we describe a certain natural construction (going back to Murray and von Neumann), which, given a pair (α, A), yields a noncommutative dynamical system (β, B) 'extending' (α, A) and such that $\beta \in \mathrm{Aut}(\mathsf{B})$ is inner. Assume that A is faithfully and non-degenerately represented on a Hilbert space H and consider the Hilbert space $\ell^2(\mathbb{Z}) \otimes \mathsf{H}$, which we can identify with the space of all sequences $(\xi_n)_{n\in\mathbb{Z}}$ of vectors in H for which $\sum_{n\in\mathbb{Z}} \|\xi_n\|^2 < \infty$. For each $a \in \mathsf{A}$ consider the following operator $\iota(a)$ on $\ell^2(\mathbb{Z}) \otimes \mathsf{H}$:

$$\iota(a)((\xi_n)_{n\in\mathbb{Z}}) = (\alpha^{-n}(a)\xi_n)_{n\in\mathbb{Z}}, \quad (\xi_n)_{n\in\mathbb{Z}} \in \ell^2(\mathbb{Z}) \otimes \mathsf{H}.$$

It is easy to check that each $\iota(a)$ is bounded and further that the map $\iota : \mathsf{A} \to B(\ell^2(\mathbb{Z}) \otimes \mathsf{H})$ is a unital injective representation of A. Let further $u \in B(\ell^2(\mathbb{Z}) \otimes \mathsf{H})$ denote the amplification of the two-sided shift map on $\ell^2(\mathbb{Z})$:

$$u((\xi_n)_{n\in\mathbb{Z}}) = (\xi_{n+1})_{n\in\mathbb{Z}}, \quad (\xi_n)_{n\in\mathbb{Z}} \in \ell^2(\mathbb{Z}) \otimes \mathsf{H};$$

The operator u is of course a unitary. The C^*-algebra generated by $\pi(\mathsf{A})$ and u in $B(\ell^2(\mathbb{Z}) \otimes \mathsf{H})$ is called the *crossed product of* A *by* α and is usually denoted $\mathsf{A} \rtimes_\alpha \mathbb{Z}$. The notation reflects the fact that the automorphism α can be always viewed as an action of $\mathbb{Z}$. As one can check (using some straightforward norm computations), the construction above does not depend (up to an isomorphism) of the choice of the initial representation of A on H. We will always use the notation $\iota : \mathsf{A} \to \mathsf{A} \rtimes_\alpha \mathbb{Z}$ to describe the embedding of A into the relevant crossed product and write u for the corresponding unitary. Sometimes it will be more convenient to identify the sequence $(\xi_n)_{n\in\mathbb{Z}}$ with the (Hilbert space norm) convergent series $\sum_{n\in\mathbb{Z}} \delta_n \otimes \xi_n$ and work directly with individual vectors $\delta_k \otimes \xi \in \ell^2(\mathbb{Z}) \otimes \mathsf{H}$ ($k \in \mathbb{Z}, \xi \in \mathsf{H}$).

Exercise 2.4.1 Formalize the remark on the independence of the construction of $\mathsf{A} \rtimes_\alpha \mathbb{Z}$ of the choice of the original representation of A. Note that we can also speak of $\iota : \mathsf{A} \to \mathsf{A} \rtimes_\alpha \mathbb{Z}$ and $u \in \mathsf{A} \rtimes_\alpha \mathbb{Z}$ without the risk of confusion.

Exercise 2.4.2 Can you identify the C^*-subalgebra of $\mathsf{A} \rtimes_\alpha \mathbb{Z}$ generated by the operator u (see Section 2.1.5)? In particular, what is $\mathbb{C} \rtimes_{\mathrm{id}_\mathbb{C}} \mathbb{Z}$?

Consider the following computation ($a \in \mathsf{A}, \xi \in \mathsf{H}, k \in \mathbb{Z}$) :

$$\begin{aligned}(u^*\iota(a)u)(\delta_k \otimes \xi) &= (u^*\iota(a))(\delta_{k-1} \otimes \xi) = u^*(\delta_{k-1} \otimes \alpha^{1-k}(a)\xi) \\ &= \delta_k \otimes \alpha^{1-k}(a)\xi = \delta_k \otimes \alpha^{-k}(\alpha(a))\xi \\ &= \iota(\alpha(a))(\delta_k \otimes \xi).\end{aligned}$$

Thus, identifying A with $\iota(\mathsf{A})$, we observe that the inner automorphism $\mathrm{Ad}\, u$ of $\mathsf{A} \rtimes_\alpha \mathbb{Z}$ 'extends' α:

$$(\mathrm{Ad}\, u) \circ \iota = \iota \circ \alpha. \tag{2.4.1}$$

It can be further shown that $\mathsf{A} \rtimes_\alpha \mathbb{Z}$ is in fact the universal C^*-algebra containing a copy of A and a unitary u 'implementing' the action α. This is related to the fact that $\mathbb{Z}$ is *amenable* – see the comments after Theorem 2.4.2.

Remark 2.4.1 Equality (2.4.1) implies in particular that the 'normally ordered' set $\{\iota(a)u^l : a \in \mathsf{A}, l \in \mathbb{Z}\}$ is linearly dense in $\mathsf{A} \rtimes_\alpha \mathbb{Z}$.

Exercise 2.4.3 Let (α, A) be a noncommutative dynamical system with $\alpha \in \mathrm{Aut}(\mathsf{A})$ and let $\beta \in \mathrm{Aut}(\mathsf{A})$ pointwise commute with α (that is, $\beta(\alpha(a)) = \alpha(\beta(a))$ for $a \in \mathsf{A}$). Use the universality of the crossed product (as alluded to above) to show that there exists a unique automorphism $\widetilde{\beta}$ of $\mathsf{A} \rtimes_\alpha \mathbb{Z}$ such that

(i) $\widetilde{\beta} \circ \iota = \iota \circ \beta$;

(ii) $\widetilde{\beta}(u) = u$.

Observe that $\widetilde{\alpha} = \operatorname{Ad} u$.

2.4.2 The Voiculescu entropy computations for the maps extended to crossed products

One of the first natural questions regarding the Voiculescu entropy of a given automorphism is whether it does not change under passing to its natural 'inner extension' (that is, the extension to the crossed product constructed as in Exercise 2.4.3). The following result shows that this is indeed the case.

Theorem 2.4.2 ([Vo]) Let A be a nuclear C^*-algebra, let $\alpha \in \operatorname{Aut}(\mathsf{A})$ and let $\beta \in \operatorname{Aut}(\mathsf{A})$ pointwise commute with α. Then $\mathsf{A} \rtimes_\alpha \mathbb{Z}$ is nuclear and moreover $\operatorname{ht} \beta = \operatorname{ht} \widetilde{\beta}$.

Proof The fact that $\mathsf{A} \rtimes_\alpha \mathbb{Z}$ is nuclear is a by-product of the proof below; therefore, we will not focus on this part (although formally we should first prove it so that we can formally speak of $\operatorname{ht} \widetilde{\beta}$).

We first fix $n \in \mathbb{N}$ and construct unital completely positive maps $\psi_n : \mathsf{A} \rtimes_\alpha \mathbb{Z} \to M_n(\mathsf{A})$ and $\phi_n : M_n(\mathsf{A}) \to \mathsf{A} \rtimes_\alpha \mathbb{Z}$. The first of them is defined in the following way: let $P_n : \ell^2(\mathbb{Z}) \otimes \mathsf{H} \to \ell^2(\mathbb{Z}) \otimes \mathsf{H}$ be the orthogonal projection onto $\overline{\operatorname{Lin}}\{\delta_k \otimes \xi : k = 1, \ldots, n, \xi \in \mathsf{H}\}$. Note that we can (and will) view P_n as a map from $\ell^2(\mathbb{Z}) \otimes \mathsf{H}$ to $\mathbb{C}^n \otimes \mathsf{H}$. As $\mathsf{A} \rtimes_\alpha \mathbb{Z} \subset B(\ell^2(\mathbb{Z}) \otimes \mathsf{H})$ we can consider the (unital, completely positive) map

$$\psi_n(x) = P_n x P_n^*, \quad x \in \mathsf{A} \rtimes_\alpha \mathbb{Z}.$$

A priori ψ_n takes values in $B(\mathbb{C}^n \otimes \mathsf{H}) \approx M_n(B(\mathsf{H}))$, but consider the following computation ($a \in \mathsf{A}, \xi \in \mathsf{H}, l \in \mathbb{Z}, k \in \{1, \ldots, n\}$)

$$\psi_n(\iota(a)u^l)(\delta_k \otimes \xi) = P_n \iota(a)(\delta_{k-l} \otimes \xi) = P_n(\delta_{k-l} \otimes \alpha^{l-k}(a)(\xi))$$

and the latter is equal either to 0 (if $k - l \notin \{1, \ldots, n\}$) or to $\delta_{k-l} \otimes \alpha^{l-k}(a)(\xi)$ (if $k - l \in \{1, \ldots, n\}$). Thus, using the picture of matrix units as before we see that

$$\psi_n(\iota(a)u^l) = \sum_{\substack{k=1 \\ 1\leq k-l\leq n}}^{n} e_{k-l,k} \otimes \alpha^{l-k}(a). \tag{2.4.2}$$

In particular $\psi_n(\iota(a)u^l) \in M_n(\mathsf{A})$ and Remark 2.4.1 implies that $\psi_n : \mathsf{A} \rtimes_\alpha \mathbb{Z} \to M_n(\mathsf{A})$. The map $\phi_n : M_n(\mathsf{A}) \to \mathsf{A} \rtimes_\alpha \mathbb{Z}$ is defined by the following explicit formula:

$$\phi_n(e_{r,s} \otimes a) = \frac{1}{n} u^{-r}\iota(a)u^s = \frac{1}{n}\iota(\alpha^r(a))u^{s-r},\ r,s = 1,\dots,n, a \in \mathsf{A}. \tag{2.4.3}$$

Unitality of ψ_n is easy to check. Complete positivity follows from the fact that ψ_n can be written as a composition of completely positive maps $\iota^{(n)}$ and $x \mapsto V^*xV$, where $x \in M_n \otimes B(\ell^2(\mathbb{Z}) \otimes \mathsf{H}) \approx B(\mathbb{C}^n \otimes \ell^2(\mathbb{Z}) \otimes \mathsf{H})$ and $V \in B(\ell^2(\mathbb{Z}) \otimes \mathsf{H}; \mathbb{C}^n \otimes \ell^2(\mathbb{Z}) \otimes \mathsf{H})$, with $V = [u^1, \cdots, u^n]$. The defining formulas (2.4.2) and (2.4.3) and the definition of $\widetilde{\beta}$ allow us to check easily that we have

$$\psi_n \circ \widetilde{\beta} = \beta^{(n)} \circ \psi_n, \quad \widetilde{\beta} \circ \phi_n = \phi_n \circ \beta^{(n)}. \tag{2.4.4}$$

Further for $a \in \mathsf{A}$ and $l \in \mathbb{Z}$ we have

$$\begin{aligned}
\phi_n\left(\psi_n(\iota(a)u^l)\right) &= \phi_n\left(\sum_{\substack{k=1 \\ 1\leq k-l\leq n}}^{n} e_{k-l,k} \otimes \alpha^{l-k}(a)\right) \\
&= \sum_{\substack{k-1 \\ 1\leq k-l\leq n}}^{n} \phi_n\left(e_{k-l,k} \otimes \alpha^{l-k}(a)\right) \\
&= \frac{1}{n}\sum_{\substack{k=1 \\ 1\leq k-l\leq n}}^{n} \iota(\alpha^{k-l}(\alpha^{l-k}(a)))u^l \\
&= \frac{1}{n}\sum_{\substack{k=1 \\ 1\leq k-l\leq n}}^{n} \iota(a)u^l = \frac{n-|l|}{n}\iota(a)u^l.
\end{aligned}$$

This fact together with Remark 2.4.1 implies (remember that unital completely positive maps are automatically contractive) that for each $x \in \mathsf{A} \rtimes_\alpha \mathbb{Z}$

$$\lim_{n\to\infty} \|\phi_n(\psi_n(x)) - x\| = 0.$$

We are now ready to begin the main part of the proof. Monotonicity of the Voiculescu entropy with respect to passing to invariant subalgebras and an observation that (A,β) and $(\iota(\mathsf{A}),\widetilde{\beta}|_{\iota(\mathsf{A})})$ are isomorphic noncommutative dynamical systems implies that $\operatorname{ht}\beta \leq \operatorname{ht}\widetilde{\beta}$. Thus, it suffices to show the reverse inequality. Let $\Omega \subset\subset \mathsf{A} \rtimes_\alpha \mathbb{Z}$ and let $N \in \mathbb{N}$ be such that for each $x \in \Omega$ there is

$$\|\phi_N(\psi_N(x)) - x\| \leq \frac{\epsilon}{2}.$$

Let $\mathsf{Z} = \phi_N(\Omega) \subset\subset M_N(\mathsf{A})$. Consider now $k \in \mathbb{N}$ and the set $\mathsf{Z}^{(k)} = \bigcup_{j=0}^{k-1}(\beta^{(N)})^j(\mathsf{Z})$. Let $(\gamma,\rho,M_n) \in CPA(\mathsf{Z}^{(k)},\frac{\epsilon}{2})$. Then our claim is that $(\phi_N \circ \gamma, \rho \circ \psi_N, M_n) \in CPA(\Omega^{(k)},\frac{\epsilon}{2})$, where of course $\Omega^{(k)} = \bigcup_{j=0}^{k-1}(\beta^{(N)})^j(\Omega)$. Indeed, let $j \in \{0,\ldots,k-1\}$ and $x \in \Omega$. Then

$$\begin{aligned}
&\|(\phi_N \circ \gamma \circ \rho \circ \psi_N)(\widetilde{\beta}^j(x)) - \widetilde{\beta}^j(x)\| \\
&\qquad \leq \| (\phi_N \circ \gamma \circ \rho \circ \psi_N)(\widetilde{\beta}^j(x)) - (\phi_N \circ \psi_N \circ \widetilde{\beta}^j)(x)\| \\
&\qquad\quad + \|(\phi_N \circ \psi_N)(\widetilde{\beta}^j(x)) - \widetilde{\beta}^j(x)\| \\
&\qquad \leq \| \left(\phi_N \circ \gamma \circ \rho \circ (\beta^{(N)})^j\right)(z) - \left(\phi_N \circ (\beta^{(N)})^j\right)(z)\| \\
&\qquad\quad + \|\widetilde{\beta}^j\left((\phi_N \circ \psi_N)(x) - x\right)\|,
\end{aligned}$$

where $z = \psi_N(x)$, so that $v := (\beta^{(N)})^j(z) \in \mathsf{Z}^{(k)}$. Thus further,

$$\begin{aligned}
&\|(\phi_N \circ \gamma \circ \rho \circ \psi_N)(\widetilde{\beta}^j(x)) - \widetilde{\beta}^j(x)\| \\
&\qquad \leq \phi_N((\gamma \circ \rho)(v) - v\| + \|(\phi_N \circ \psi_N)(x) - x\| \leq \frac{\epsilon}{2} + \frac{\epsilon}{2} = \epsilon.
\end{aligned}$$

Finally, we see that

$$\begin{aligned}
\limsup_{k\to\infty}\left(\frac{1}{k}\log \operatorname{rcp}(\Omega^{(k)},\epsilon)\right) &\leq \limsup_{k\to\infty}\left(\frac{1}{k}\log \operatorname{rcp}(\mathsf{Z}^{(k)},\frac{\epsilon}{2})\right) \\
&\leq \operatorname{ht}\beta^{(N)} = \operatorname{ht}\beta,
\end{aligned}$$

where the last equality follows from Exercise 2.2.8. As $\Omega \subset\subset \mathsf{A} \rtimes_\alpha \mathbb{Z}$ and $\epsilon > 0$ were arbitrary, the proof is finished. □

The key ingredient of the proof above is the 'approximate invariance' of the set $\{1,\ldots,N\}$ under shifting by l for large N.

Such approximately invariant sets in an arbitrary group are called *Følner sets*; their existence is equivalent to the *amenability* of the group in question, that is, the existence of a state on the algebra $\ell^\infty(\Gamma)$ that is invariant under the left and right translations. In particular, as mentioned before, $\mathbb{Z}$ is amenable. The argument above is just the tip of an iceberg of connections between amenability and finite-dimensional approximations in the theory of C^*-algebras. Several examples of such connections can be found in [BrO]; in fact Theorem 2.4.2 can be generalized to the crossed products by actions of arbitrary locally compact amenable groups (Chapter 8 of [NS]) and even discrete amenable *quantum groups* ([SZ3]).

Of course the simplest and in a sense the most interesting example of the application of the above theorem is obtained by putting $\beta = \alpha$.

Corollary 2.4.3 Let A be a nuclear C^*-algebra, let $\alpha \in \mathrm{Aut}(\mathsf{A})$ and let $\mathrm{Ad}\, u \in \mathrm{Aut}(\mathsf{A} \rtimes_\alpha \mathbb{Z})$ be the canonical automorphism extending α. Then $\mathrm{ht}\, \alpha = \mathrm{ht}\, \mathrm{Ad}\, u$.

The corollary means that computing the Voiculescu entropy of inner automorphisms is as complicated as that of general automorphisms (as each automorphism can be 'made inner' via the crossed product extension and this procedure does not change the Voiculescu entropy).

2.5 Quantum 'Measurable' Dynamical Systems and Classical Ergodic Theorems

For the last two sections of these lectures, we switch our attention from the entropy considerations and topological dynamics to ergodic theorems and measurable dynamics.

2.5.1 Measurable dynamical systems and individual ergodic theorem

So far, we talked mainly about topological dynamical systems. Classically, an equally important class is that of so-called *measurable dynamical systems*, that is, pairs $((Y,\mu),T)$, where (Y,μ) is a probability space and $T : (Y,\mu) \to (Y,\mu)$ is a *measure preserving invertible transformation* (this means that both T and T^{-1}

are measurable and for each measurable $A \subset Y$ there is $\mu(T(A)) = \mu(A)$). In this context we can also speak of entropy (as mentioned before), but we want to focus on the ergodic-type results, describing the long-time properties of the system.

The following two results are specific instances of the two cornerstones of the classical ergodic theory: von Neumann's mean ergodic theorem and Birkhoff's individual ergodic theorem.

Theorem 2.5.1 (von Neumann) Let $((Y, \mu), T)$ be a measurable dynamical system. Then the formula

$$U_T(f) = f \circ T, \quad f \in L^2(Y, \mu)$$

defines a unitary operator on $L^2(Y, \mu)$. Write $\mathrm{Fix}\, U_T := \{f \in L^2(Y, \mu) : U_T(f) = f\}$ and let $P \in B(L^2(Y, \mu))$ be the orthogonal projection onto $\mathrm{Fix}\, U_T$. Then for each $f \in L^2(Y, \mu)$

$$\frac{1}{n} \sum_{k=0}^{n-1} (U_T)^k(f) \overset{n\to\infty}{\longrightarrow} Pf \tag{2.5.1}$$

in L^2-norm.

Theorem 2.5.2 (Birkhoff) Let $((Y, \mu), T)$ be a measurable dynamical system. Then for each $f \in L^1(Y, \mu)$ the sequence $(\frac{1}{n} \sum_{k=0}^{n-1} f \circ T^k)_{n=1}^{\infty}$ is almost surely convergent (to a function in $L^1(Y, \mu)$).

For a very accessible presentation of the proofs of these classical results and far-reaching extensions (mostly in the classical context) we refer to the book [Kre]. Here we just mention that the first theorem consists of two parts: the first is the construction of the unitary operator U_T (sometimes called the *Koopman representation* of $((Y, \mu), T)$), and the second is a general statement valid for all Hilbert space unitaries (in fact all Hilbert space contractions).

Later we will discuss certain extensions of the ergodic theorems involving the simultaneous study of several iterations of a given map T.

2.5.2 GNS construction and the passage from topological to measurable noncommutative dynamical systems

In Sections 2.1–2.4 we investigated certain properties of topological noncommutative dynamical systems. Now we will describe the fundamental in quantum dynamics (and more generally in the theory of operator algebras) Gelfand–Naimark–Segal construction. It will allow us to build in a natural way noncommutative 'measurable' dynamical systems.

Theorem 2.5.3 (Gelfand, Naimark, Segal) Let A be a C^*-algebra and let $\omega \in \mathsf{A}^*$ be a state. Then there exists a Hilbert space H_ω, a unital representation $\pi_\omega : \mathsf{A} \to B(\mathsf{H}_\omega)$ and a vector $\Omega \in \mathsf{H}_\omega$ such that

$$\overline{\mathrm{Lin}}\, \pi_\omega(\mathsf{A})\Omega = \mathsf{H}_\omega,$$

$$\omega(a) = \langle \Omega, \pi_\omega(a)\Omega \rangle, \quad a \in \mathsf{A}.$$

The triple $(\pi_\omega, \mathsf{H}_\omega, \Omega)$ is unique up to a unitary isomorphism: by this we mean that if $(\pi', \mathsf{H}', \Omega')$ is another triple as above, then there exists a unique unitary operator $U : \mathsf{H}_\omega \to \mathsf{H}'$ such that $U(\Omega) = \Omega'$ and $\pi'(a) = U\pi_\omega(a)U^*$ for all $a \in \mathsf{A}$. If ω is faithful, then the representation $\pi_\omega : \mathsf{A} \to B(\mathsf{H}_\omega)$ is faithful.

What is the connection of the last result with measurable dynamical systems? It turns out that a natural class of operator algebras corresponding to the classical algebras of the L^∞-type is the class of *von Neumann algebras*.

Definition 2.5.4 Let H be a Hilbert space. For any vectors $\xi, \eta \in \mathsf{H}$ define a seminorm $p_{\xi,\eta}$ on $B(\mathsf{H})$ by the formula

$$p_{\xi,\eta}(T) = |\langle \xi, T\eta \rangle|, \quad T \in B(\mathsf{H}).$$

The *weak operator topology* on $B(\mathsf{H})$ is the (locally convex) topology determined by the family of seminorms $\{p_{\xi,\eta} : \xi, \eta \in B(\mathsf{H})\}$. In other words, a net $(T_i)_{i\in I}$ converges to T in the weak operator topology if and only if for all $\xi, \eta \in \mathsf{H}$ the net $(p_{\xi,\eta}(T_i))_{i\in I}$ converges to $p_{\xi,\eta}(T)$.

Definition 2.5.5 A von Neumann algebra is a unital *-subalgebra of $B(\mathsf{H})$ closed in the weak operator topology.

von Neumann algebras can be equivalently, and in a sense more naturally, defined as unital *-subalgebras closed in the *ultraweak (σ-weak) operator topology*. The latter, in general finer than the weak operator topology (but coinciding with it on bounded subsets), turns out to be the most important (apart from the one coming from the norm) topology in the study of von Neumann algebras. More information on the latter topology and the proofs of the theorems stated below can be found in the monograph [StZ]. Note that each von Neumann algebra is in particular a (concrete) unital C^*-algebra.

Theorem 2.5.6 (von Neumann) Let $\mathcal{A}$ be a *-subalgebra of $B(\mathsf{H})$ and let $\mathcal{A}' = \{T \in B(\mathsf{H}) : \forall_{a\in\mathcal{A}}\, Ta = aT\}$, $\mathcal{A}'' = \{T \in B(\mathsf{H}) : \forall_{a'\in\mathcal{A}'}\, Ta' = a'T\}$. The following are equivalent:

(i) $\mathcal{A}$ is a von Neumann algebra;

(ii) $\mathcal{A} = \mathcal{A}''$.

Note that it follows from the above result that if $\mathcal{A}$ is a unital *-subalgebra of $B(\mathsf{H})$ then it is weak-operator dense in its double commutant $\mathcal{A}''$. A natural example of a von Neumann algebra, similar to what we have already seen in Section 2.1, is the *von Neumann algebra of a discrete group* Γ, $\mathrm{VN}(\Gamma)$. It is defined as the weak-operator-closure of the reduced group C^*-algebra $C_r^*(\Gamma)$; in other words, $\mathrm{VN}(\Gamma) = C_r^*(\Gamma)'' \subset B(\ell^2(\Gamma))$.

Theorem 2.5.7 (GNS construction – commutative case) Let X be a compact space, μ – a regular probability measure on X, $\mathsf{A} = C(X)$. Let $\omega \in S(\mathsf{A})$ be given by integration with respect to the measure μ. Then the GNS construction for the pair (A, ω) leads to the triple $(\pi_\mu, L^2(X,\mu), 1_X)$ for which the von Neumann algebra $\pi_\mu(C(X))''$ is isometrically isomorphic to the algebra $L^\infty(X,\mu)$ (the algebra of the essentially bounded functions on (X,μ), equipped with the essential supremum norm).

Thus, the algebra of the form $L^\infty(X,\mu)$ provides an example of a (commutative) von Neumann algebra. Its ultraweak operator topology is the weak*-topology induced by the duality $L^\infty(X,\mu) = (L^1(X,\mu))^*$. In fact this is a general result: Sakai's Theorem ([Sa]) characterizes abstract von Neumann algebras as those C^*-algebras that are dual Banach spaces: $\mathsf{M} = (\mathsf{M}_*)^*$, where M_* is the so-called *predual* of M (and can be thought of as a generalized L^1-space). Using the standard functional-analytic trick of embedding a Banach space into its second dual, we often view the predual of M as the subspace of the dual of M: $\mathsf{M}_* \subset \mathsf{M}^*$. The identification of M with the dual of M_* implies in particular that M_*, viewed as a subspace of M^*, separates points of M.

A linear map T between von Neumann algebras is called *normal* if it is continuous with respect to relevant ultraweak topologies. Example of a normal map is given by the map of the form $\mathsf{M} \ni y \mapsto x^*yx$, where $x \in B(\mathsf{H})$ is such that $x^*yx \in \mathsf{M}$ for each $y \in \mathsf{M}$. Using the identification from the above paragraph and a general fact regarding the elements of the Banach space bidual we see that normal functionals in M^* are precisely those which belong to M_*. Examples of normal functionals on $\mathsf{M} \subset B(\mathsf{H})$ are given by *vector functionals*, that is, maps of the form $\mathsf{M} \ni x \mapsto \langle \xi, x\eta \rangle$, where $\xi, \eta \in \mathsf{H}$. The uniqueness property of the GNS construction (up to a unitary equivalence) leads to an easy proof of the following result.

Proposition 2.5.8 Let A be a C^*-algebra with a faithful state $\omega \in \mathsf{A}^*$ and let $\alpha \in \mathrm{Aut}(\mathsf{A})$ be such that $\omega \circ \alpha = \omega$. Then there exists a (unique) normal automorphism $\tilde{\alpha} \in \mathrm{Aut}(\pi_\omega(\mathsf{A})'')$ such that

$$\tilde{\alpha}(\pi_\omega(a)) = \pi_\omega(\alpha(a)), \quad a \in \mathsf{A}.$$

Exercise 2.5.1 Prove the last proposition using information contained in the notes. Can you see the similarities/differences between this extension of α and the one appearing in the crossed product considerations?

It is easy to observe that in the situation described in the last proposition there exists a (unique) normal state $\widetilde{\omega} \in S(\pi(\mathsf{A})'')$ such that for $a \in \mathsf{A}$ there is $\widetilde{\omega}(\pi_\omega(a)) = \omega(a)$. Moreover, the automorphism $\tilde{\alpha}$ preserves $\widetilde{\omega}$.

Exercise 2.5.2 Prove the above statements.

Generally, if M is a von Neumann algebra and $\omega \in S(\mathsf{M})$ is a normal state, we will write $\alpha \in \mathrm{Aut}(\mathsf{M}, \omega)$ to express the fact that α is a normal automorphism of M preserving the state ω: $\omega = \omega \circ \alpha$. Note that the GNS construction can be also applied to von Neumann algebras. In that case in a sense the construction changes very little (as the resulting von Neumann algebra is isomorphic to the original one), but often still remains very useful. We state some of its aspects in the following proposition.

Proposition 2.5.9 Let M be a von Neumann algebra and $\omega \in S(\mathsf{M})$ a faithful normal state. Then the GNS representation π_ω

is a (normal) isomorphism of the von Neumann algebra M and $\pi_\omega(\mathsf{M})'' = \pi_\omega(\mathsf{M})$. Moreover, the collection of functionals $\{M \ni x \to \langle \xi, \pi_\omega(x)\eta \rangle : \xi, \eta \in \mathsf{H}_\omega\}$ spans M_*.

We finish this section by returning once again to the example of the group von Neumann algebra for a discrete group Γ. Consider the vector functional on $\mathrm{VN}(\Gamma)$ given by the formula

$$\tau(x) = \langle \delta_e, x\delta_e \rangle, \quad x \in \mathrm{VN}(\Gamma), \tag{2.5.2}$$

where e denotes the neutral element of Γ.

Proposition 2.5.10 The vector functional τ defined in (2.5.2) is a faithful normal trace on $\mathrm{VN}(\Gamma)$, called the *canonical trace* on $\mathrm{VN}(\Gamma)$.

Proof The fact that τ is a state is easy to verify, as $\tau(x^*x) = \|x\delta_e\|^2$ for any $x \in \mathrm{VN}(\Gamma)$. Its normality is a consequence of one of the facts stated before Proposition 2.5.8. To show faithfulness we need first to note that elements of $\mathrm{VN}(\Gamma)$ commute with the right shifts $\{r_\gamma : \gamma \in \Gamma\} \subset B(\ell^2(\Gamma))$, where

$$r_\gamma(\delta_{\gamma'}) = \delta_{\gamma'\gamma^{-1}}, \quad \gamma, \gamma' \in \Gamma$$

(compare the above formula with formula (2.1.2)). As $\mathrm{VN}(\Gamma) = C_r^*(\Gamma)''$, it suffices to show that elements of $C_r^*(\Gamma)$ commute with the right shifts; by linearity and continuity it further suffices to show that left shifts commute with right ones, that is, $\lambda_\gamma r_{\gamma'} = r_{\gamma'}\lambda_\gamma$ for all $\gamma, \gamma' \in \Gamma$. The last formula can be verified directly.

Suppose then that $x \in \mathrm{VN}(\Gamma)$ and $\tau(x^*x) = 0$. This means that $\|x\delta_e\|^2 = 0$, so $x\delta_e = 0$. However, then for any $\gamma \in \Gamma$ we have $0 = r_{\gamma^{-1}}x\delta_e = xr_{\gamma^{-1}}\delta_e = x\delta_\gamma$, so that $x = 0$.

To verify the tracial property of τ it again suffices because of linearity and continuity to check that for any $\gamma, \gamma' \in \Gamma$ we have $\tau(\lambda_\gamma\lambda_{\gamma'}) = \tau(\lambda_{\gamma'}\lambda_\gamma)$. This however is equivalent with the statement

$$\gamma\gamma' = e \Longleftrightarrow \gamma'\gamma = e,$$

which is clearly true. □

Exercise 2.5.3 Use the notation of the above proposition and denote the von Neumann algebra generated in $B(\ell^2(\Gamma))$ by the right shift operators by $\mathcal{R}(\Gamma)$. Prove, using Theorem 2.5.6 and comments after it, that

$$\mathrm{VN}(\Gamma)' = \mathcal{R}(\Gamma)$$

(some authors write $\mathcal{L}(\Gamma)$ instead of $\mathrm{VN}(\Gamma)$).

2.6 Noncommutative Ergodic Theorem of Lance; Classical and Quantum Multi Recurrence

In this last section we discuss some noncommutative extensions of the classical ergodic theorems presented in Section 2.5 and also sketch both classical and quantum results concerning so-called non-conventional, multiparameter averages.

2.6.1 Mean ergodic theorem(s) in von Neumann algebras

The convergence appearing in the mean ergodic theorem is the convergence in the Hilbert space norm, and as such still makes sense when a noncommutative version is considered. We make it more specific in what follows: note that if we start from the framework described in Theorem 2.5.7 then the usual L^2-norm of an element $f \in \pi_\omega(C(X))'' = L^\infty(X,\mu)$ can be expressed by the formula $\|f\|_2 = (\omega(f^*f))^{\frac{1}{2}}$. Similarly, it is not too difficult to verify that if (M,ω) is a von Neumann algebra with a faithful normal state, then M equipped with a new norm: $\|x\|_2 := (\omega(x^*x))^{\frac{1}{2}}$ $(x \in \mathsf{M})$ becomes a pre-Hilbert space. You may note that the proof is based on the crucial inequality valid for all states on C^*-algebras, the *Kadison–Schwarz inequality* (which is also a fundamental tool in the GNS construction): if A is a C^*-algebra and $\omega \in S(\mathsf{A})$ then

$$|\omega(x^*y)| \leq (\omega(xx^*))^{\frac{1}{2}}(\omega(y^*y))^{\frac{1}{2}}, \quad x,y \in \mathsf{A}.$$

The respective completion is usually denoted as $L^2(\mathsf{M},\omega)$ and called the *noncommutative L^2-space associated with* (M,ω). Given $\alpha \in \mathrm{Aut}(\mathsf{M},\omega)$ we can then easily verify that the map $x \mapsto \alpha(x), x \in \mathsf{M}$ extends to a unitary operator U_α on $L^2(\mathsf{M},\omega)$ (see the proof of Theorem 2.6.1 below). This unitary can be viewed as the noncommutative Koopman construction for α; it satisfies a suitably modified counterpart of the formula (2.5.1) of

Theorem 2.5.1. Note that the above statements can be viewed as another way of phrasing of the GNS construction.

There is another possible approach to generalizing the mean ergodic theorem to the context of von Neumann algebras, which can be traced back to the paper [KS]. We describe it here briefly in a simplified version to illustrate the notions and techniques introduced in Section 2.5.

Theorem 2.6.1 Let M be a von Neumann algebra, let $\omega \in \mathsf{M}^*$ be a faithful normal tracial state and let $\alpha \in \mathrm{Aut}(\mathsf{M}, \omega)$. Define $\mathrm{Fix}\,\alpha = \{m \in \mathsf{M} : \alpha(x) = x\}$. Then there exists a contractive idempotent map Φ from M onto $\mathrm{Fix}\,\alpha$, which has the following properties:

(i) Φ is positive;

(ii) $\omega \circ \Phi = \omega$;

(iii) for all $\gamma \in \mathsf{M}_*$ and $x \in \mathsf{M}$

$$\lim_{n\to\infty} \frac{1}{n} \sum_{k=0}^{n-1} \gamma(\alpha^k(x)) = \gamma(\Phi(x)). \tag{2.6.1}$$

Proof Let $(\pi_\omega, \mathsf{H}_\omega, \Omega)$ be the GNS construction for (M, ω) and define for each $x \in \mathsf{M}$

$$U(\pi_\omega(x)\Omega) = \pi_\omega(\alpha(x))\Omega.$$

Then U is an isometry on the space $\pi_\omega(\mathsf{M})\Omega \subset \mathsf{H}_\omega$: for $x, y \in \mathsf{M}$

$$\begin{aligned}\langle U(\pi_\omega(x)\Omega), U(\pi_\omega(y)\Omega)\rangle &= \langle \pi_\omega(\alpha(x))\Omega, \pi_\omega(\alpha(y))\Omega\rangle \\ &= \langle \Omega, \pi_\omega(\alpha(x)^*\alpha(y))\Omega\rangle = \omega(\alpha(x^*y)) \\ &= \omega(x^*y) = \langle \pi_\omega(x)\Omega, \pi_\omega(y)\Omega\rangle.\end{aligned}$$

The fact that $\pi_\omega(\mathsf{M})\Omega$ is dense in H_ω implies that U extends to an isometry on the whole Hilbert space; using the aforementioned density and invertibility of α we conclude that U is in fact unitary. Thus, (the abstract version of) the von Neumann's mean ergodic theorem implies that for every $\xi \in \mathsf{H}_\omega$

$$\frac{1}{n} \sum_{k=0}^{n-1} U^k(\xi) \overset{n\to\infty}{\longrightarrow} P\xi, \tag{2.6.2}$$

where P is a projection in $B(\mathsf{H}_\omega)$.

Fix now a normal functional $\gamma \in \mathsf{M}_* \subset \mathsf{M}^*$ and define for each $n \in \mathbb{N}, x \in \mathsf{M}$

$$s_n(\gamma, x) = \frac{1}{n}\sum_{k=0}^{n-1} \gamma \circ \alpha^k(x).$$

Assume that there exist $y, z \in \mathsf{M}$ such that $\gamma(x) = \langle \pi_\omega(y)\Omega, \pi_\omega(x)\pi_\omega(z)\Omega\rangle$ for $x \in \mathsf{M}$. Then we have for each $k \in \mathbb{N}_0$

$$\begin{aligned}\gamma(\alpha^k(x)) &= \langle \pi_\omega(y)\Omega, \pi_\omega(\alpha^k(x))\pi_\omega(z)\Omega\rangle = \omega(y^*\alpha^k(x)z) = \omega(zy^*\alpha^k(x)) \\ &= \langle \pi_\omega(yz^*)\Omega, \pi_\omega(\alpha^k(x))\Omega\rangle = \langle \pi_\omega(yz^*)\Omega, U^k(\pi_\omega(x)\Omega)\rangle,\end{aligned}$$

where we used the fact that ω is tracial. Thus for this particular γ using the equality (2.6.2) we obtain that

$$s_n(\gamma, x) = \left\langle \pi_\omega(yz^*), \frac{1}{n}\sum_{k=0}^{n-1} U^k(\pi_\omega(x)\Omega)\right\rangle \overset{n\to\infty}{\longrightarrow} \langle \pi_\omega(yz^*), P(\pi_\omega(x)\Omega)\rangle.$$

As it is easy to see that $\|s_n(\gamma, x)\| \leq \|\gamma\|\|x\|$ for any $\gamma \in \mathsf{M}_*$ and $x \in \mathsf{M}$, linearity of the map $\gamma \mapsto s_n(\gamma, x)$, density of $\pi_\omega(\mathsf{M})\Omega$ in H_ω and the last statement of Proposition 2.5.9 implies that the sequence of complex numbers $(s_n(\gamma, x))_{n=1}^\infty$ is in fact convergent for any $\gamma \in \mathsf{M}_*$, $x \in \mathsf{M}$. Denote the relevant limit by $\tilde{s}(\gamma, x)$, fix for the moment $x \in \mathsf{M}$ and consider the map

$$\mathsf{M}_* \ni \gamma \mapsto \tilde{s}(\gamma, x).$$

It is easy to check using again the properties listed in the last paragraph that this map is a bounded linear functional on M_* (of norm not greater than $\|x\|$). As $\mathsf{M} = (\mathsf{M}_*)^*$ there exists $m_x \in \mathsf{M}$ uniquely determined by the equality

$$\gamma(m_x) = \tilde{s}(\gamma, x), \quad \gamma \in \mathsf{M}_*. \tag{2.6.3}$$

Put $\Phi(x) = m_x$, $x \in \mathsf{M}$ and note that formulas (2.6.1) and (2.6.3) coincide. The fact that so-defined map Φ is linear and contractive is easy to verify.

To show that $\Phi(\mathsf{M}) \subset \mathrm{Fix}\,\alpha$ it suffices to note that for every $x \in \mathsf{M}$ and $\gamma \in \mathsf{M}_*$ we have $\gamma(\Phi(x)) = \gamma(\alpha(\Phi(x))$, or in other words that $\tilde{s}(\gamma, x) = \tilde{s}(\gamma \circ \alpha, x)$. The last formula follows easily once we write down explicit limits. On the other hand, if $x \in \mathrm{Fix}\,\alpha$ then

$\widetilde{s}(\gamma, x) = \gamma(x)$ and thus $x = \Phi(x)$. These two facts combined show that Φ is an idempotent map, with the image equal to Fix α. The fact that Φ is positive is a consequence of the fact that all the operations involved in the construction are positivity-preserving (the reader is invited to verify the details).

Finally for any $x \in \mathsf{M}$ we have $\omega(\Phi(x)) = \widetilde{s}(\omega, x) = \omega(x)$. This ends the proof. □

The theorem remains valid without the assumption on traciality of ω: the general case can be reduced to the tracial one via the Tomita–Takesaki theory (see for example Chapters VII–VIII of [Ta$_2$]). The map Φ constructed above can be naturally interpreted as the *noncommutative conditional expectation* onto Fix α preserving the state ω. It can be further shown to be completely positive and normal (in fact each contractive idempotent E from a C^*-algebra A onto a C^*-subalgebra B is automatically completely positive, and is a *module map* with respect to a smaller subalgebra: $E(b_1 a b_2) = b_1 E(a) b_2$ for all $a \in \mathsf{A}, b_1, b_2 \in \mathsf{B}$ – see Theorem 3.4 [Ta$_1$] – it only proves positivity but complete positivity can be shown very similarly). It is also determined uniquely as a contractive projection onto Fix α preserving the state ω, which follows from the next exercise.

Exercise 2.6.1 Use the information contained in brackets in the last paragraph to prove that if A is a C^*-algebra with a faithful state $\omega \in S(\mathsf{A})$ and B is a C^*-subalgebra of A then the norm-one idempotent E from A onto B preserving ω, if it exists, must be unique. Would this be true if we do not assume that E preserves ω? Can you find a (commutative, finite-dimensional) counterexample?

The last part of the above exercise partially explains the role of the state ω in Theorem 2.6.1.

2.6.2 Almost uniform convergence in von Neumann algebras

Extending the almost sure convergence of the individual ergodic theorem to the noncommutative context is more problematic – we first need to find a way of expressing it without using the 'point' picture.

Theorem 2.6.2 (Egorov Theorem) A sequence $(f_n)_{n=1}^{\infty}$ of measurable functions on a probability space (Y, μ) is convergent pointwise (almost surely) to a function f if and only if for each $\epsilon > 0$ there exists $Z \subset Y$ such that $\mu(Y \setminus Z) < \epsilon$ and $(f_n|_Z)_{n=1}^{\infty}$ is uniformly convergent to $f|_Z$.

The above theorem effectively replaces the almost sure convergence with the uniform convergence on 'large enough' subsets. To pass to the noncommutative setting, we first replace sets by characteristic functions (that is, orthogonal projections in the algebra $L^\infty(Y, \mu)$) and note that $f_n|_Z = f_n \chi_Z$. This leads to the following definition, due to Segal.

Definition 2.6.3 Let M be a von Neumann algebra with a normal faithful state $\omega \in \mathsf{M}^*$. We say that a sequence $(x_n)_{n=1}^{\infty}$ of elements in M converges to $x \in \mathsf{M}$ almost uniformly, if for each $\epsilon > 0$ there exists an orthogonal projection $p \in \mathsf{M}$ such that

$$\omega(p^\perp) < \epsilon \quad \text{and} \quad \lim_{n\to\infty} \|(x_n - x)p\| = 0.$$

Note that in fact we could also consider a so-called *bilateral almost uniform* convergence, replacing above $\|(x_n - x)p\|$ by $\|p(x_n - x)p\|$.

2.6.3 Lance's noncommutative individual ergodic theorem and some comments on its proof

We are now ready to present a noncommutative counterpart of Birkhoff's individual ergodic theorem, owing to Lance.

Theorem 2.6.4 ([La]) Let M be a von Neumann algebra, let $\omega \in \mathsf{M}^*$ be a faithful normal state and let $\alpha \in \mathrm{Aut}(\mathsf{M}, \omega)$. Then for every $x \in \mathsf{M}$ the sequence $(\frac{1}{n}\sum_{k=0}^{n-1} \alpha^k(x))_{n=1}^{\infty}$ is convergent almost uniformly to an element of M.

Sketch of the proof The proof of the above theorem, similarly as the proof of Birkhoff's Theorem, consists of two main parts. The first one roughly corresponds to showing that Theorem 2.6.1, or rather its non-tracial generalization, holds. The second requires proving a version of a so-called Maximal Lemma, which in this case looks as follows: for any $\epsilon > 0$ and $x \in \mathsf{M}_+$ such that $\omega(x) = \epsilon$ there exists $y \in \mathsf{M}$ such that

$$\|y\| \leq 2, \quad \omega(y) \leq 4\epsilon^{\frac{1}{2}}$$

and

$$\frac{1}{n}\sum_{k=0}^{n-1}\alpha^k(x) \leq y, \quad n \in \mathbb{N}.$$

The proof of the Maximal Lemma is based on a detailed analysis of properties of certain affine functions defined on a compact convex subset of a locally convex linear topological space. □

Classical Birkhoff's theorem is usually formulated for L^1-functions (and this is how we stated it in Theorem 2.5.2), and not only for essentially bounded ones. In fact, Lance's theorem also admits generalizations to 'noncommutative L^p-spaces', but even stating them reaches beyond the content of our lectures. Interested readers should consult the paper [JX] and the bibliography therein.

2.6.4 Classical multirecurrence

Recent years have brought a rapid development in the study of the so-called *multirecurrence* or *'multi-ergodic'* theorems. Below we present the appropriate definitions in the von Neumann algebraic language.

Definition 2.6.5 Let M be a von Neumann with a faithful normal state ω and let $\alpha \in \mathrm{Aut}(\mathsf{M},\omega)$. We say that $(\mathsf{M},\omega,\alpha)$ enjoys order k convergence in norm if for any $x_1,\ldots,x_{k-1} \in \mathsf{M}$ the sequence $(y_n)_{n=1}^{\infty}$ defined by the formula

$$y_n := \frac{1}{n}\sum_{j=1}^{n}(\alpha^j(x_1))(\alpha^{2j}(x_2))\cdots(\alpha^{(k-1)j}(x_{k-1})), \quad n \in \mathbb{N},$$

converges in $L^2(\mathsf{M},\omega)$ – in other words it satisfies the Cauchy condition in $\|\cdot\|_2$ -norm, that is,

$$\lim_{n,m\to\infty}\omega((y_n - y_m)^*(y_n - y_m)) = 0.$$

For classical measurable dynamical systems ($\mathsf{M} = L^{\infty}(Y,\mu)$, $\alpha = \alpha_T$) the latter condition means simply that the sequence $(y_n)_{n=1}^{\infty}$ converges in $L^2(Y,\mu)$. Using Lance's ergodic theorem one can show that each triple $(\mathsf{M},\omega,\alpha)$ enjoys order-2 convergence in norm. An important theorem of Host and Kra from [HK] can be reformulated in the following way.

Theorem 2.6.6 Let $((Y,\mu),T)$ be a measurable dynamical system, let $\mathsf{M} = L^\infty(Y,\mu)$, let $\omega \in S(\mathsf{M})$ be induced by the measure μ and let $\alpha \in \mathrm{Aut}(\mathsf{M},\omega)$ be induced by T. Then for each $k \in \mathbb{N}$ the triple $(\mathsf{M},\omega,\alpha)$ enjoys order k convergence in norm.

The origins of the above result can be traced to the fundamental work of Furstenberg. An accessible survey of this and other classical multi recurrence results, together with the description of fascinating connections to combinatorics and number theory can be found in [Kra].

2.6.5 Noncommutative extensions and counterexamples due to Austin, Eisner, and Tao

A recent paper [AET] contains, apart from several positive results, a construction of an example which shows that Theorem 2.6.6 cannot be fully generalized to noncommutative systems.

The algebra M considered in the required example is the group von Neumann algebra of a certain group Γ defined by generators and relations; the trace τ is the canonical trace and the automorphism α is induced from an automorphism of Γ (via the construction described in Lemma 2.6.8 below). We first introduce another definition.

Definition 2.6.7 An automorphism α of a von Neumann algebra M is said to be *ergodic* if its fixed point space $\mathrm{Fix}\,\alpha$ consists only of the scalar multiples of the unit.

Lemma 2.6.8 Let Γ be a discrete group and let $\psi : \Gamma \to \Gamma$ be an automorphism. Then there exists a (unique) normal automorphism $\alpha_\psi : \mathrm{VN}(\Gamma) \to \mathrm{VN}(\Gamma)$ such that

$$\alpha_\psi(\lambda_\gamma) = \lambda_{\psi(\gamma)}, \quad \gamma \in \Gamma. \tag{2.6.4}$$

The automorphism α_ψ preserves the canonical trace τ. If, moreover, for each $k \in \mathbb{Z} \setminus \{0\}$ and $\gamma \in \Gamma \setminus \{e\}$ we have $\psi^k(\gamma) \neq \gamma$ (that is, no power of ψ has non-trivial fixed points), α_ψ is ergodic.

Proof Let Γ and ψ be as above. As ψ is bijective, we can define a unitary U_ψ on $\ell^2(\Gamma)$ by the continuous linear extension of the formula

$$U_\psi \delta_\gamma = \delta_{\psi^{-1}(\gamma)}, \quad \gamma \in \Gamma;$$

it is easy to verify that

$$U_\psi{}^* \delta_\gamma = \delta_{\psi(\gamma)}, \quad \gamma \in \Gamma.$$

Further for any $\gamma, \gamma' \in \Gamma$

$$\begin{aligned} U_\psi{}^* \lambda_\gamma U_\psi \delta_{\gamma'} &= U_\psi{}^* \lambda_\gamma \delta_{\psi^{-1}(\gamma')} = U_\psi{}^* \delta_{\gamma\psi^{-1}(\gamma')} \\ &= \delta_{\psi(\gamma\psi^{-1}(\gamma'))} = \delta_{\psi(\gamma)\gamma'} = \lambda_{\psi(\gamma)} \delta_{\gamma'}. \end{aligned}$$

It follows from the above that the automorphism of $B(\ell^2(\Gamma))$ given by $x \mapsto U_\psi{}^* x U_\psi$ restricts to a normal automorphism of $\mathrm{VN}(\Gamma)$ (this can be shown either via the topological argument or using the identification of the commutant of $\mathrm{VN}(\Gamma)$ provided in Exercise 2.5.3). Denote the resulting restriction by α_ψ and note that it satisfies the equality (2.6.4).

The fact that α_ψ preserves τ follows from the equalities $U_\psi \delta_e = U_\psi{}^* \delta_e = \delta_e$.

It remains to establish ergodicity of α_ψ under the assumption that no power of ψ has non-trivial fixed points. Begin from the following observation: for any $\gamma \in \Gamma$ and $\xi, \eta \in \ell^2(\Gamma)$ we have:

$$\lim_{n\to\infty} \langle \xi, \alpha_\psi^n(\lambda_\gamma)\eta \rangle \overset{n\to\infty}{\longrightarrow} \langle \xi, \tau(\lambda_\gamma)\eta \rangle.$$

Indeed, as all $\alpha_\psi^n(\lambda_\gamma)$ are unitaries (therefore, in particular contractions), it suffices to verify it for $\xi = \lambda_{\gamma'}$, $\eta = \lambda_{\gamma''}$ for $\gamma', \gamma'' \in \Gamma$. Then however the expression on the left of the displayed limit relation is equal to 1 if $\psi^n(\gamma') = \gamma\psi^n(\gamma'')$ and zero otherwise; in other words, it is equal to 1 only if $\gamma = \psi^n(\gamma'(\gamma'')^{-1})$. If $\gamma = e$, then this holds if and only if $\gamma' = \gamma''$. On the other hand if $\gamma \neq e$, then for the equality above to hold we cannot have $\gamma' = \gamma''$. To finish the analysis of the left-hand side it suffices to note that an orbit of any non-zero element of Γ under ψ cannot contain repetitions (as if $\psi^n(t) = \psi^m(t)$ then $\psi^n(t)$ is a fixed point for ψ^{m-n}); thus, if $\gamma \neq e$ then the limit on the left is always equal to 0. On the other hand the expression on the right is easily seen to be equal to 1 if $\gamma = e$ and $\gamma' = \gamma''$ and to 0 otherwise.

Suppose now that $\xi \in \ell^2(\Gamma)$ is a fixed point for the unitary U_ψ. Then for any $n \in \mathbb{N}$ and $\gamma \neq e$

$$\langle \xi, \lambda_\gamma \xi \rangle = \langle U_\psi^n \xi, \lambda_\gamma U_\psi^n \xi \rangle = \langle \xi, \alpha^n(\lambda_\gamma)\xi \rangle \overset{n\to\infty}{\longrightarrow} \langle \xi, \tau(\lambda_\gamma)\xi \rangle = 0.$$

As γ was arbitrary, this implies that ξ is supported at most one point of Γ, and this point is easily seen to be necessary equal to e (as ψ has no non-trivial fixed points). Thus $\xi \in \mathbb{C}\delta_e$.

Finally, suppose that $x \in \mathrm{VN}(\Gamma)$ is a fixed point for α_ψ. Then

$$U_\psi{}^* x\delta_e = U_\psi{}^* x U_\psi \delta_e = \alpha_\psi(x)\delta_e = x\delta_e.$$

Thus, $x\delta_e$ is a fixed vector for $U_\psi{}^*$, so also for U_ψ. This means that $x\delta_e = \mu\delta_e$ for some $\mu \in \mathbb{C}$. However, x commutes with all the right shifts (see the proof of Proposition 2.5.10), so we can deduce that $x\delta_\gamma = \mu\delta_\gamma$ for any $\gamma \in \Gamma$. Thus $x = \mu 1$. □

Now the fact that we can construct counterexamples to a potential generalization of Theorem 2.6.6 to noncommutative systems is based on the following purely group-theoretic result.

Lemma 2.6.9 ([AET]) Let $A \subset \mathbb{Z}$. Then there exists a group Γ with elements $\gamma_0, \gamma_1, \gamma_2, \gamma_3 \in \Gamma$ and an automorphism ψ of Γ such that no power of ψ has non-trivial fixed points and if $r \in \mathbb{Z}$ then

$$\gamma_0 T^r(\gamma_1) T^{2r}(\gamma_2) T^{2r}(\gamma_3) = e \text{ if and only if } r \in A.$$

Proof See Appendix B to [AET]. □

The group in question is obviously highly noncommutative. Its construction is based on the concept of *square groups* and uses certain highly non-trivial combinatorial and geometric techniques.

We can now state precisely the result we mentioned in the first line of this section.

Theorem 2.6.10 ([AET]) There exists a von Neumann algebra M with a faithful normal tracial state τ and an ergodic automorphism $\alpha \in \mathrm{Aut}(\mathsf{M}, \tau)$ such that $(\mathsf{M}, \tau, \alpha)$ does not enjoy order 4 convergence in norm.

Exercise 2.6.2 Deduce Theorem 2.6.10 from Lemmas 2.6.8 and 2.6.9.

On the other hand, note that to ensure that $(\mathsf{M}, \omega, \alpha)$ enjoys order k convergence in norm for any $k \in \mathbb{N}$ it suffices to assume that ω is a tracial state and the system is *asymptotically commutative*, that is, for all $x, y \in \mathsf{M}$

$$\lim_{n\to\infty} \frac{1}{n} \sum_{k=0}^{n-1} \|\alpha^k(x)y - y\alpha^k(x)\|_2 = 0.$$

The last result is also proved in [AET].

2.6.6 Final remarks

The above notes of course touch only a few aspects of the theory of noncommutative – or 'not necessarily commutative' – dynamical systems. This theory in a sense 'contains' the classical theory, but, as we had a chance to see, extending even basic concepts or theorems related to 'standard' dynamical systems to the noncommutative setting requires usually completely new tools, mostly coming from the theory of operator algebras (but also from group theory, matrix analysis, combinatorics, and so on), and often also forces additional assumptions. Recent decades have seen similar developments of *noncommutative probability* and *noncommutative geometry*.

References

[AET] T. Austin, T. Eisner and T. Tao, Nonconventional ergodic averages and multiple recurrence for von Neumann dynamical systems, *Pacific J. Math.* **250** (2011), no. 1, 1–60.

[AF] R. Alicki and M. Fannes, "Quantum dynamical systems," Oxford University Press, Oxford, 2001.

[BG] F. Boca and P. Goldstein, Topological entropy for the canonical endomorphisms of Cuntz–Krieger algebras, *Bull. London Math. Soc.* **32** (2000), no. 3, 345–352.

[Br] N. Brown, Topological entropy in exact C^*-algebras, *Math. Ann.* **314** (1999), no. 2, 347–367.

[BrO] N. Brown and N. Ozawa, "C^*-Algebras and finite dimensional approximations", Graduate Studies in Mathematics, 88. American Mathematical Society, Providence, RI, 2008.

[CHS_1] R. Conti, J.H. Hong and W. Szymański, Endomorphisms of the Cuntz algebras, *Banach Center Publ.* **96** (2011), 81–97.

[CHS_2] R. Conti, J.H. Hong and W. Szymański, Endomorphisms of graph algebras, *J. Funct. Anal.* **263** (2012), no. 9, 2529–2554.

[CRS] R. Conti, M. Rørdam and W. Szymański, Endomorphisms of O_n which preserve the canonical UHF-subalgebra, *J. Funct. Anal.* **259** (2009), no. 3, 602–617.

[Ch] M. Choda, Entropy of Cuntz's canonical endomorphism, *Pacific J. Math.* **190** (1999), no. 2, 235–245.

[Cu_1] J. Cuntz, Simple C^*-algebras generated by isometries. *Comm. Math. Phys.* **57** (1977), no. 2, 173–185.

[Cu_2] J. Cuntz, Automorphisms of certain simple C^*-algebras, in 'Quantum fields—algebras, processes' (Proc. Sympos., Univ. Bielefeld, Bielefeld, 1978), pp. 187–196, Springer, 1980.

[DF] T. Downarowicz and B. Frej, Measure-theoretic and topological entropy of operators on function spaces, *Ergodic Th. Dynam. Systems* **25** (2005), no. 2, 455–481.

[He] H. Helson, "The spectral theorem", Lecture Notes in Mathematics **1227**. Springer–Verlag, Berlin, 1986.

[Hi] F. Hiai, Entropy for canonical shifts and strong amenability, *Internat. J. Math.* **6** (1995), no. 3, 381–396.

[HK] B. Host and B. Kra, Nonconventional ergodic averages and nilmanifolds, *Ann. Math.* **161** (2005), no. 1, 397–488.

[JX] M. Junge and Q. Xu, Noncommutative maximal ergodic theorems, *J. Amer. Math. Soc.* **20** (2007), 385–439.

[Ka] K. Kawamura, Polynomial endomorphisms of the Cuntz algebras arising from permutations. I. General theory., *Lett. Math. Phys.* **71** (2005), no. 2, 149–158.

[Ke] D. Kerr, Entropy and induced dynamics on state spaces, *Geom. Funct. Anal.* **14** (2004), no. 3, 575–594.

[KL] D. Kerr and H. Li, Dynamical entropy in Banach spaces, *Invent. Math.* **162** (2005), 649–686.

[KS] I. Kovács and J. Szücs, Ergodic type theorems in von Neumann algebras, *Acta Sci. Math. (Szeged)* **27** (1966), 233–246.

[Kra] B. Kra, From combinatorics to ergodic theory and back again, in *International Congress of Mathematicians, Vol.III*, pp.57–76, Eur.Math.Soc., Zürich, 2006. *Available at:* http://www.math.northwestern.edu/ kra/papers.html.

[Kre] U. Krengel, "Ergodic theorems," de Gruyter Studies in Mathematics, 6. Walter de Gruyter & Co., Berlin, 1985.

[La] C.E. Lance, Ergodic theorems for convex sets and operator algebras, *Invent. Math.* **37** (1976), no. 3, 201–214.

[Mu] G.J. Murphy, "C^*-algebras and operator theory," Academic Press, Inc., Boston, 1990.

[Na] M.G. Nadkarni, "Spectral theory of dynamical systems", Reprint of the 1998 original. *Texts and Readings in Mathematics*, **15**. Hindustan Book Agency, New Delhi, 2011.

[NS] S.Neshveyev and E.Størmer, "Dynamical entropy in operator algebras," Ergebnisse der Mathematik und ihrer Grenzgebiete. 3. Folge. A Series of Modern Surveys in Mathematics, 50. Springer-Verlag, Berlin, 2006.

[Ra] I. Raeburn, "Graph C^*-algebras", CBMS Regional Conference Series in Mathematics, 103, Providence, RI, 2005.

[Sa] S. Sakai, "C^*-algebras and W^*-algebras," reprint of the 1971 edition, Classics in Mathematics. Springer–Verlag, Berlin, 1998.

[Sk_1] A. Skalski, On automorphisms of C^*-algebras whose Voiculescu entropy is genuinely noncommutative, *Ergodic Th. Dynam. Systems*, **31** (2011), 951–954.

[Sk_2] A. Skalski, Noncommutative topological entropy of endomorphisms of Cuntz algebras II, *Publ. Res. Inst. Math. Sci.* **47** (2011), no. 4, 887–896.

[SZ_1] A. Skalski and J. Zacharias, Entropy of shifts in higher rank graph C^*-algebras, *Houston Journal of Mathematics* **34** (2008), no.1, 269–282.

[SZ_2] A. Skalski and J. Zacharias, Noncommutative topological entropy of endomorphisms of Cuntz algebras, *Lett. Math. Phys.* **86** (2008), no. 2–3, 115–134.

[SZ_3] A. Skalski and J. Zacharias, Approximation properties and entropy estimates for crossed products by actions of amenable discrete quantum groups, *J. Lond. Math. Soc.* **82** (2010), no. 1, 184–202.

[StZ] S. Stratila and L. Zsido, "Lectures on von Neumann algebras," Abacus Press, Tunbridge Wells, 1979.

[Ta_1] M. Takesaki, "Theory of operator algebras. I", Springer-Verlag, New York–Heidelberg, 1979.

[Ta_2] M. Takesaki, "Theory of operator algebras. II". Encyclopaedia of Mathematical Sciences, **125**. Springer-Verlag, Berlin, 2003

[Vo] D. Voiculescu, Dynamical approximation entropies and topological entropy in operator algebras, *Comm. Math. Phys.* **170** (1995), no. 2, 249–281.

[Wa] P. Walters, "An introduction to ergodic theory," Graduate Texts in Mathematics, 79. *Springer–Verlag, New York–Berlin,* 1982.

[Wo] S.L. Woronowicz, Compact quantum groups, in *Symétries Quantiques, Les Houches, Session LXIV, 1995*, pp. 845–884.

Index